Pennsylvania German Farms, Gardens, and Seeds

Landis Valley in Four Centuries

Irwin Richman

Schiffer Publishing Ltd

4880 Lower Valley Road, Atglen, PA 19310 USA

Other Schiffer Books by Irwin Richman
German Architecture in America.
Pennsylvania German Arts: More Than Hearts, Parrots, & Tulips.

Other Schiffer Books on Related Subjects
American Barns. Stanley Schuler.
The Farm: An American Living Portrait. Joan and David Hagan.
Fraktur Folk Art and Family. Corinne & Russell Earnest.
Victorian Gardens. Caroline Holmes.

Library of Congress Cataloging-in-Publication Data

Richman, Irwin.
Pennsylvania German farms, gardens, and seeds : Landis Valley in four centuries / Irwin Richman.
p. cm.
Includes bibliographical references.
ISBN 0-7643-2546-9 (pbk.)
1. Pennsylvania Dutch—Pennsylvania—Landis Valley—History. 2. Family farms—Pennsylvania—Landis Valley—History. 3. Gardens—Pennsylvania—Landis Valley—History. 4. Seeds—Collection and preservation—Pennsylvania—Landis Valley. I. Title.

F160.G3R54 2007
974.8'1500431—dc22
2006017957

Designed by Mark David Bowyer
Type set in University Roman Bd BT / Souvenir Lt BT

ISBN: 0-7643-2546-9
Printed in China

Published by Schiffer Publishing Ltd.
4880 Lower Valley Road
Atglen, PA 19310
Phone: (610) 593-1777; Fax: (610) 593-2002
E-mail: Info@schifferbooks.com

Contents

Acknowledgments

Since retiring from the Pennsylvania State University at Harrisburg in 2003, I have been working with the Landis Valley Associates holding the grand title of Director of Research and Development of the Heirloom Seed Project. One of my joyous duties has been to write this book, all the proceeds of which go to the non-profit Landis Valley Associates, whose mission it is to help maintain and develop the Landis Valley Museum in Manheim Township, Lancaster County, Pennsylvania, a unique site where agriculture and horticulture has been practiced from the 1700s to today: now into a fourth century. Landis Valley, whenever possible, is my armature for telling the story of *Pennsylvania German Farms, Gardens and Seeds.*

Any project relies on the help and support and contributions of many people. Special thanks go to Dr. Howard Sachs as Associate Dean for Research and Development at Penn State Harrisburg for urging me to do "A German Garden Book." Stephen S. Miller, the Director of the Landis Valley Museum, has been a major facilitator. I am also proud to claim him as a former graduate student. He has helped to shape the modern Landis Valley Museum. He is an avid gardener, a professional leader, and a humane administrator. My other colleagues at Landis Valley have also been helpful. Thank you, Marilyn Monath, Donna Horst, Bruce Bomberger, Tim Essig, Will Morrow, Joyce Perkinson, Cindy Reedy, Joe Schott, and Trish Frey. Of very special note is Michael Emery, Landis Valley Museum Educator and Volunteer Coordinator, who is a font of knowledge about the Pennsylvania Germans and an avid collector of books, paper, and photographic memorabilia. He has shared many items with me and has ferreted out others. His fingerprints are clearly visible in the book. Craig Benner, also on the Landis Valley staff, provided fine, original photography. Special thanks also go to Nadine Steinmetz (AKA Nettie May Landis) for her inspiration and her photographs.

Thanks also go to the institution and individuals who have provided other illustrations for the work. All are acknowledged in the captions. Especially helpful were the staffs at the Heritage Center of Lancaster County, The Moravian Archives, Historic Salem, Inc., The Schwenkfelder Library and Heritage Center, The Pennsylvania Academy of the Natural Sciences, The American Folk Art Museum, The Pennsylvania Dutch Tourist Bureau, The Lancaster-York Heritage District, Oregon Dairy, Kreider Farms, and, of course, the Pennsylvania Historical and Museum Commission, which owns and administers the Landis Valley Museum.

Being technologically challenged, like Blanche DuBois in Tennessee Williams' play "A Streetcar Named Desire", I rely on the kindness of others. Reference librarians consider me quaint. Those at Penn State Harrisburg apparently like me around as a relic. As always, they have been extremely helpful. Thank you Greg Crawford (now the, still helpful, Director of the Library), Alan Mays, Eric Delozier, and Glen McGuigan. At the office Mike Emery always helps me with computer issues. I do not know, however, if I could continue to function in this increasingly complex world without my wife, Dr. M. Susan Richman, mathematician and retired university dean. She is my major enabler and she has voluntarily transcribed the manuscript from my inelegant scrawl into useful and improved typescript. She, thankfully, is touched with the work ethic of her Pennsylvania German forebears. Her mother, Florence C. Steigerwalt, and my late father-in-law, the Reverend Eugene O. Steigerwalt, have been (and are) informants and supporters.

I write for a public that wants and enjoys information about my adoptive culture. I also write for future generations, especially Sue and my descendents. Our son, Dr. Alexander E. Richman, is a mathematician who has recently married Elana R. Pressman, an environmental policy person. Our younger son, Dr., Dr. Joshua S. Richman (he has both an M.D. and a Ph.D.) and his wife Kristin Graham Richman are the parents of our grandchildren, Benjamin David (5) and Zoë Elizabeth Anne (3). Both Benjamin and Zoë like to help their parents and eat the heirloom crops they grow, currently, in diaspora, in Alabama.

The Cottage
Bainbridge, Pennsylvania
2005

I. Farming and Gardening: Meet the Pennsylvania Germans

The Expo Center at the Lancaster Host Resort is huge, white, and fittingly barn-like. On a November evening in 2004 over 600 people are attending the 28th annual Agriculture Industry Banquet, sponsored by the Lancaster Chamber of Commerce and Industry. Many are farmers; more are in agribusiness. Some are certainly named Landis. Most are dressed in Middle America Rotary meeting dress-up. There are some variations. Among the women, white prayer veils – sheer cap-like garments worn by Mennonite women – are visible. Mary Martin, whose Narvon, Pennsylvania, farm is being recognized as the 2004 "Lancaster County Century Farm," is there in a long dress, some glittery ornaments, and a white prayer veil. Her husband, John David Martin, looks prosperous in banker blue with a red tie. His great-great-great-grandfather bought their farm in 1854. The couple's twenty-eight-year-old son, Jonathan, now runs it. The generations share a large, now sub-divided, Federal style farmhouse built in 1834. The farm has seventy acres of land and the Martins milk seventy-six Holsteins. Over the years survival has not always been easy but, in true Pennsylvania German devotion to hard work, determined faith, and a tradition of good land stewardship, the family persevered and has often prospered. During the Great Depression, when John David's grandfather ran the farm, his cows were stricken with tuberculosis and had to be destroyed. In 1994, Bovine viral diarrhea or BVD again almost ended the Martins' farming operations. One in four of their cows were stricken and the others had to be sold to prevent the disease's spread. Mrs. Martin recalls, "We cleaned the barn and left it set for a year." In the meantime, her husband got a job in a cement plant with no plans to resume dairying, but their children missed the cows and urged them to get back into the business. "They said it was not right to have no cows here, to not go into the barn with Dad."

Caring for their land has always been important. John David's father, Paul, was one of the first farmers in Lancaster County's Caernavon Township to plant feed crops in contour strips in order to limit erosion, and in the late 1970s the Martins built a water retention pond with the same goal in mind. Despite rising land values, in 1996 the family permanently preserved their farm in an agricultural land bank. John David recalls that he had fallen in love with the farm in the 1950s while driving his father's S.C. Case tractor over its acres. Mary notes, "It's a good farm life raising children on a farm." She proudly adds, "The farm was here before the Civil War."

Corn ripens on the Myers family's 265 acre farm near the town of Willow Street in Lancaster County. The 2001 Century Farm Award winner has been in the family for six generations. In addition to field crops, the Myers raise vegetables and flowers. About 2/3 of their harvest is sold to local grocery stores, universities, and retirement communities. The remainder is sold at their produce stand where "sunflowers, zinnias, and dahlias augment the vegetable crops." *Lancaster-York Heritage Region*

The Martin family is warmly applauded by the crowd in the Expo Center. Elsewhere in the room are couples in Plain Garb – mostly more traditional Mennonites and River Brethern. In other words, the audience is typical of an agricultural event in the area. Flip through the program and read the names of the 2004 Lancaster Chamber Agriculture Committee: Beam, Benner, Hurst, Peachey, Risser, Rohrer, Shellenberger, and Wolgemuth: German all. The George C. Delp award "... is presented annually to a resident of Lancaster County who has made a significant personal contribution to preserving and enhancing agriculture in Lancaster County." Past winners bear surnames Funk, Newswanger, Rutt, Wenger, Schwartz, Shirk, and Rider. More Germans. This year's winner is egg farmer-industrialist Paul Sauder, Jr. Another German. The author had never before attended an agricultural meeting of this scope with a formal program, and now suddenly it dawned on him that he is totally surrounded by Pennsylvania Germans, most of whose families had been farming and gardening in the region for parts of four centuries. Here they all were. Some were sectarians in garb, but most were normatively dressed folks whose immediate "Dutchiness" is only identifiable by their surnames – or occasionally, also, by an accent, usually slight, occasionally pronounced.

John David and Mary Martin were the Century Farm Award winners in 2004. Their family has owned their Narvon, Lancaster County, dairy farm since 1854 when his great-great-great-grandfather, Jacob Martin, bought what was a portion of the older Pool Forge Estate. Their large Federal-style farmhouse was built in 1838. Mary holds a nailbox, a family heirloom that dates to 1855. *Lancaster New Era, Marty Heisey, Photographer*

The rooted nature of the Germans can also be seen in the Shirk Family of Goodville, whose history of land ownership and active farming almost makes the Century Farming Martin family seem like newcomers. They are 2004's newest inductees into the Pennsylvania Bicentennial Farm Program. The Shirk farm is one of forty in the state (including three others in Lancaster County) that has met the Commonwealth's program requirements, calling for continuous ownership by one family of a parcel of land with a minimum size of ten acres for at least 200 years, producing crops, and lived on permanently by at least one member of the ownership family. The three-generation Shirk family represents the eighth, ninth, and tenth generations to live on their land.

The Shirks have traced ownership of their land to the Penn family and the farm went through six owners before being acquired in 1758 by Mathias Stauffer, one of Bill Shirk's ancestors. Bill's grandmother, Alice Stauffer, inherited the house five generations later – and she poetically named the farm "Silver Valley" because of the silvery dewy mornings. Grace met her husband, Moses Shirk, in 1941 when she was signing up men for the draft and he was in her line. After the war Bill ran the farm, including caring for sixteen to twenty dairy cows and Grace taught at neighboring schools. Like many German farm families, the Stauffer-Shirk family has been blessed with longevity. When Bill and his wife, Catherine, gave a surprise 50th Wedding Anniversary Party for his parents, Grace and Moses, the couple learned that two other couples who had lived in their old farm house, Grace's parents and grandparents, had also celebrated their golden anniversaries. Caring deeply for the land, as do the Martins, the present day Shirks have also preserved their sixty-six-acre farm in perpetuity when they sold development rights to the Commonwealth. Reflecting the new reality of many contemporary farm owners in the region, Bill laments that he cannot afford to farm fulltime. Instead he grows corn, hay, soybeans, and wheat part-time, having given up dairying and works fulltime at CNH, a farm machinery manufacturer in New Holland. Very involved in his community, he is also a volunteer fireman.

The Shirk Family of Goodville, Lancaster County, pose in front of the 1864 farmhouse on their "Silver Valley Farm," which has been in the family since 1758. It is recognized as a Bicentennial Farm by the United States Department of Agriculture. The children in the front row are, no doubt, in training to keep the family tradition going. *Lancaster Intelligencer Journal, Vinny Tennis, Photographer*

Even in conservative Dutchland, the modern, more diverse, age is reflected in the family genetic pool. Bill's wife, Catherine, is of Scandinavian descent – though with strong farming roots. One of the couple's sons, Jim, and his family live on the farm in a new house built on an acre plot carved from their homestead, and he works in agribusiness, for Wenger Feeds, and helps with the farm. Very proud of his heritage, Jim pushed for his family's involvement with the Bicentennial Farm Program. However, not all farm children choose to remain on the farm. Jim's brother, John, lives in Minnesota, working for Lockheed Martin. His mother observes, "John is much more an urban person. Where he got that I don't know." But it happens in the best of farm families. The Martins' son has followed his father into farming, but their daughters, Janessa Fisher and Jacinda Martin, have left the century farm – and farming – as did John Shirk.

The commitment of Pennsylvania German farmers to the maintenance of their land was further highlighted at the 54th annual banquet of the Lancaster County Conservation District. Annual awards, Certificates of Excellence, were presented to six farmers – all of Dutch Origin. The "Outstanding Cooperator Award" was awarded to Scott and Herb Kreider and Sons Dairy in East Drumore. The Kreider operation is massive. Four generations farm 645 acres and milk about 400 cattle. First establishing a conservation plan in 1974, "... the Kreiders have completed more than 8 1/2 miles of crop terraces that prevent erosion." As a result of their work, "... streams that receive runoff from the Kreiders farms are designated high-quality cold-water fisheries."

The six Certificate winners were (1) Jere and Donna Bollinger of Lititz, who raise steers and 106,000 broiler chickens, (2) Lamar Moyer, who cares for 84,000 broilers on his twelve-acre farm near Marietta, (3) Manheim's John and Sharon Brubaker, who raise 112,000 broilers, (4) Keith Martin of Stevens, who raises 58,000 broilers, (5) Gregory Herr, who maintains 122,000 broilers at the Herr Family Farm near Manheim, and (6) Robert and Regina Zimmerman, who raise 58,000 broilers on their Reinholds area farm. Also honored with the District's Educator Award was Jeff Stoltzfus, an adult agriculture instructor for Eastern Lancaster County School District, whose "... nontraditional teaching allows him to work with many Plain-sect farms ... where he promotes streambank fencing, stream-side buffers, cover crops and other conservation practices." These individuals and families, along with thousands of others, rejoice in their Dutch heritage.

The German presence in Pennsylvania began at an extraordinarily early date, in 1683, when a group of German Mennonites created a settlement near Philadelphia which would later be called "Germantown." Germantown, which was eventually absorbed into the City of Philadelphia in the nineteenth century, would have a long and distinguished history as an independent community and as a city neighborhood, as part of William Penn's vision for a "Greene Country Towne." William Penn (1644-1718), the social and political radical, the son of Admiral Sir William Penn (1621-1670), a distinguished naval officer knighted by King Charles II, was a Quaker. Despite his family status, the younger William was subjected to harassment against which his wealth allowed him to defend himself. His less well-to-do brethren were not as lucky. Penn was determined to help them to find sanctuary. As a man of his time, he looked toward the new Eden – America.

The Penn family had been strong supporters of the monarchy and had helped finance Charles II's restoration to the throne in 1660. Admiral Penn had loaned the monarch a huge sum of money. Cash-starved Charles II owed the Admiral's estate 16,000 pounds. Poor in money, Charles was rich in claims to territory in America. To satisfy his debt, he made William Penn's dream come true. In 1681, he granted a multimillion-acre colony, which he named "Pennsylvania" in honor of the Admiral, to William Penn as proprietor. A proprietor could prosper from his colony only as much as the colony would prosper. First and foremost, the colony needed a transplanted European population. The new colony had fewer than a thousand Swedes and Hollanders, a few hundred Englishmen, and thousands of Native Americans (Indians), whose numbers had been greatly decimated because of smallpox and measles inadvertently introduced by the first European settlements. Penn encouraged Quakers to emigrate, and eventually thousands responded, but he realized these would not be enough. A shrewd early-day-real-estate-developer, Penn, through a series of writings, encouraged German Quakers and other sectarians to emigrate to his "Noble Experiment." More painfully persecuted than Quakers in England, they responded, and Germantown was founded as a beachhead for the tens of thousands of Germans, church people, and sectarians who would immigrate into Pennsylvania by 1800. It is from these seventeenth and eighteenth century immigrants that the Pennsylvania Dutch, the "Persistent Minority," are descended.

If the Pennsylvania Dutch are German in origin, why are they called "Dutch?" It is a question of semantics. The author married into an extensive, purebred Pennsylvania German family whose forebears arrived in America in the 1760s and settled in what would become Lehigh, Schuylkill, and Carbon Counties where they would live, as many continue to reside, for almost 250 years. Settlements were often rural and isolated, with few surnames. Three of Mary Susan Steigerwalt's (Richman's) grandparents were born with the Steigerwalt surname. The fourth grandparent had close Steigerwalt ancestry! The author's father-in-law, the Reverend Eugene Oscar Steigerwalt (1905-1994), a Lutheran minister, was expected, until the outbreak of World War II, to conduct one service a month in German. He gloried in being able to speak dialect German "Pennsylvania Dutch" or "Dutch" and he would proudly proclaim himself a "Dutchman," and, indeed, most Pennsylvania Germans comfortably and confidently use the "Dutch" identifier. Others felt differently.

New Hope Impressionist painter, who often celebrated the Dutch countryside in his work, Walter Emerson Baum (1884-1956), like Pastor Steigerwalt, was born of German ancestry, but he hated "Dutch" and insisted on "German," only yielding to "Dutch" during the xenophobic days of World War II. Europe had its *Plattdeutsch*, the lowlands near the sea, including the modern Dutch of Holland and its *Hochdeutsch* of modern Germany. *Deutsch* is easily corrupted into "Dutch." The term, "Pennsylvania German," is a late nineteenth century identifier that a scholarly world introduced to delineate these unique Americans. No one commonly uses the very accurate and descriptive eighteenth century term "*Pennsylfawnish Deitsch.*"

The first Germans to settle in Pennsylvania, those who established Germantown in the 1680s, built what is the oldest still-standing Mennonite meeting house in America. By the early eighteenth century, Germantown was a prosperous community of productive farms and industries, including the pioneering Rittenhouse paper mill. Germantown became famous as a center of the manufacture of woolen stockings, which were widely known in the colonies and in postcolonial America. An area of great beauty, Germantown, just six miles from the center of the colonial city of Philadelphia, would become the site of country homes of many of that city's elite. James Logan (1674-1751), William Penn's secretary who, for years, actually ran the colony, built "Stenton," which is now a museum, as is the Chew family's great house, "Cliveden," on whose lawn part of the Battle of Germantown was fought during the American Revolution.

When the railroad arrived in Germantown in the 1840s, the town became one of America's first railroad suburbs and a center of horticulture spurred on, in part, by the many gentlemen estate owners in the neighborhood. The marvelous vine, Wisteria, was named for Dr. Casper Wistar (1761-1818), whose roots were in Germantown, where his Wüster forebears settled after emigrating from the Palatinate in 1717. A noble Wisteria Vine thrives at the Wister ancestral home, "Grumblethorpe." Wister and Wistar, of course, are anglicized versions of the original family name. The Wistar family would own substantial tracts of land in central Pennsylvania, "Dutch Country," and exert great influence.

Nature's bounty was widely celebrated by German and English-descended still life artists who worked in Central Pennsylvania. Their works were especially prized among the Germans. Severin Roesen's (c.1815-c.1872) "Still Life With Fruit" was painted by the German-born artist, probably in Williamsport. John Francis (1808-1886) painted in various sites in the Susquehanna Valley. His "Melons and Fruits" is especially bountiful. Annie Snyder (1852-1927) was Pennsylvania German. "Apples and Nuts" is of extra interest because it shows American Chestnuts, a nut highly prized by the Pennsylvania Dutch. *Museum of American Art, gift of William C. Williamson by exchange and Henry S. McNeil and Henry D. Gilpin Fund; Private collection; and Schwarz Gallery, Philadelphia*

Today's Germantown is a diverse neighborhood, parts of it blighted, but it is, especially in spring, ablaze with horticultural fireworks – dogwoods, cherries, and azaleas in profusion – and its streets are mostly tree lined. It is as close to Penn's vision of a "Greene Country Towne" that modern Philadelphia has to offer.

Philadelphia itself had many German residents, and by mid-eighteenth century, German was heard as often as English in the city. Benjamin Franklin and many of his colleagues were afraid that German, rather than English, would become the colony's official language. But Philadelphia's great success as an economic hub and as, on the eve of the Revolution, the second largest city, after London, in the Empire ensured the domination of English. And while the German presence indelibly marked Philadelphia's palate (scrapple and pretzels, for example, are both German in origin), parts of rural Pennsylvania would retain their strong Germanic flavor into the twenty-first century. This rural area is what is today celebrated as "Dutch Country." It includes parts of Bucks and Montgomery Counties to the east and north of Philadelphia, as well as Lehigh, Northampton, Schuylkill, Snyder, Union, and Carbon Counties; but its heart is in the counties of South Central Pennsylvania: Berks, Lebanon, Dauphin, Lancaster, and York.

Bordering counties, including Cumberland and Perry, also have Dutch residents. The names of the counties are mostly English; "Schuylkill," however, is derived from the Holland Dutch, and "Dauphin" is French in origin. In the eighteenth century, and well into the twentieth century, these counties, with their English named cities of Reading, Lancaster, York, Allentown, Bethlehem, and Harrisburg, were largely populated with the Pennsylvania Dutch. There are still many places where people speak with a noticeable Dutch accent. The landscape in Dutch country is dotted with many smaller communities and townships whose names reflect the Germanic character of their early settlers and present inhabitants: Hamburg, Heidleburg, Strasburg, Knauers, Berlin, Reinholds, Myerstown, Wormelsdorf, and Stouchsburg are a small sample.

Just who are the Pennsylvania Dutch? The answer is surprisingly complex. First and foremost, it is important to remember that the modern country "Germany" did not exist until 1870. What is today Germany was many states and principalities, and in a world of different political boundaries, we discover that the Germans who immigrated into Pennsylvania were mostly from what is now Germany, but that some of the Pennsylvania Germans, or Pennsylvania Dutch, are not even German in origin.

The largest number of Germans came from the Palatinate, *die Pfalz*, the region near Heidelberg. Many others came from Alsace, which is, of course, now part of France. A visit to an Alsatian folk museum in France is like visiting a collection of Pennsylvania Dutch artifacts, and Alsatian cooking was widely influential on Pennsylvania Dutch cuisine. There is, incidentally, even an Alsace Township in Berks County.

Images of the fruits of the earth became very accessible to the middle class with the nineteenth century invention of chromolithography. Apple, pear, and peach star, with rose hips playing a supporting role, in a popular print widely marketed in Dutch Country. *The Cottage Collection*

Rural themes have been very popular in entertainment in the Pennsylvania Dutch region. And who could resist "Pansy the Roller-Skating Cow" as shown on a c.1940 trade card? *The Cottage Collection*

The vast majority of these pre-1800 immigrants were Lutherans and members of the German Reformed Church (now the United Church of Christ), *kirchenleute*, or "church people," who came here because they were economically stressed. They were part of a seventeenth and eighteenth century human tragedy, which forced hundreds of thousands of Germans to emigrate to Holland, to England, to Russia, and to the American colonies. They fled lands decimated by wars that raged for decades. Included among them were a smattering of Catholics and Jews. Except for the tiny number of Jews, they came primarily for economic reasons.

After the Edict of Nantes was revoked in 1685, Protestants living in France had to flee for their lives. A number of "Huguenots," as they were called, moved into the

Palatinate; and when economic conditions there deteriorated, many emigrated with their hosts to the New World where they were absorbed into the dominant Pennsylvania German culture. Other Huguenots immigrated directly from hiding places in France itself or other host countries. The "Dutch" names Zeller, Greiner, and Deturk, among others, are French Protestant in origin.

Smaller church groups and the sectarians, *sektenleute* or "Sect People," are also more cosmopolitan than one would expect. The Schwenkfelders come from Silesia, which is now Poland. Schwenkfelder culture, today, centers around Pennsburg, which predictably is near Schwenksville, Montgomery County. The Schwenkfelder Library and Heritage Center is well worth visiting. The Moravians, a German church with Reformation roots, had its beginnings in what is now the Czech Republic before moving into Germany and sending numerous members to America. Many of these Pennsylvania Dutch are really Czech or even Slovak in background. The Moravians founded several communities in Pennsylvania, the most successful of which are Lititz (named for Lidice in the Czech Republic), Nazareth, and Bethlehem. All have interesting histories and old world charm, in spots.

When most outsiders hear the phrase, "Pennsylvania Dutch," they conjure images of the Amish and the Mennonites. These archetypical Pennsylvania Dutch of popular imagination have their roots in Alsace, but most of them immigrated to America from the German-speaking cantons or states of Switzerland. A small number of Mennonites also emigrated from Holland or, like the Landises, from the Palatinate. Unlike the Lutherans and the Reformed Church members, the sectarians came here primarily for religious freedom. The worst of the persecutions these peoples endured is chronicled in a seventeenth century book, *Het bloedig toonel of Martelaers Spiegel*, or the *Martyrs Mirror*, still important to their faith. *Martyrs Mirror*, printed in Ephrata in Lancaster County in 1748, was the largest book printed in the American colonies. It contains an account of the martyrdom of Hans Landis, a progenitor of the Landis Valley Museum Landises.

Growing specialty crops for the fresh market requires pacing your output and dealing with many different plant types. This truck garden at the Myers farm in Willow Street, Lancaster County, has bush beans in the foreground, squash next, with a cabbage crop in front of the beanpoles. A member of the Myers family is in the background tending to garden chores. *Lancaster-York Heritage Region*

The largely Germanic "Dutch" culture which evolved in Pennsylvania is distinctive. Before World War II, the German government, under the Nazi regime, actively attempted to form pro *Axis Bunden* among Americans of German descent. Their efforts completely failed among the Pennsylvania Dutch who felt, and feel, absolutely no sense of loyalty to the modern presumed Fatherland. The organizers had much more success with the descendants of later immigrants. Reading, in the heart of Dutch country, attracted many late nineteenth and early twentieth century German immigrants, and it became a flourishing center of Nazi sympathy – but not among the traditional, historic residents.

Along with the Spanish of the American Southwest, the Pennsylvania Germans have nurtured and maintained a unique folk culture that has transcended centuries while interacting with and reacting to the dominant culture. The electronic media of the twentieth and twenty-first centuries, radio and television especially, have had baleful effects on the natural, organic maintenance of both of these cultures. Today, however, many scholars and laymen are devoted to the preservation of Dutch culture, which has had a surprisingly expansive effect on American culture at large. E-mail and the Web have helped as links aiding in keeping the culture alive and passing on interesting tales. Link: Abraham Lincoln was born in a log cabin; the cabin was built using German log technology. Surprised? Remember that Lincoln's grandfather grew up in Berks County, Pennsylvania – cheek to jowl with the Germans. Link: The log cabin is as American as apple pie – another German contribution to wider American culture. Pie is available in a broad range of variations and it is not always recognized for what it is. Dutch culture can also be chameleon-like.

There is an old story about a tourist stopping his car at Amos Zook's gasoline station in Wormelsdorf and asking him, "Where are the Pennsylvania Dutch?" They are all around you.

The Amish fascinate Americans and numerous books have been written about the Amish family and their community. Some of the best of these were written by the late John B. Hostetler (1918-2001). Born an Amishman, he "went gay" or "English" and became a Mennonite and an academic. His pioneering *Amish Society* has appeared in numerous editions, but there are no similar books about the church people – although there are many books about their institutions and their public lives. There are, of course, German families whose ancestors have been distinguished or famous through several generations, like the Muhlenbergs and the Wisters-Wistars, but there is very little exploration of a more typical family.

To fill a void, we now introduce you to the Steigerwalts of Central Pennsylvania – a family whose common ancestors arrived in America in 1767 and who have quietly lived, often farmed, usually gardened, and increased in America for almost 250 years.

As in other Dutch families, many Steigerwalts have a strong sense of tradition and a devotion to genealogy. Also, like many others, they hold family reunions. In common with some, they have become so numerous that they often wear nametags, occasionally identifying their line of descent. There is and was a fascination in listening to the generations born, 1880-1920, who all had, or still have, an intimate knowledge of their ancestors and who enjoyed discussing the intricacies of descent with the devotion of Talmudic scholars discussing Hebraic text.

The family's origin is in the Steigerwald region of present day Germany, a hilly part of the Palatinate, not far from Heidelberg. Many of its inhabitants took a variant of their homeland as a name, when a surname was required: Steygerwaldt, Steigerwald, Steigerwalt, etc. As a result of this practice, there are Protestant, Catholic, and Jewish Steigerwalts.

The progenitors of the Pennsylvania Steigerwalts were Johann Frederich Steigerwalt (b. 1731) and his wife, Catherine Heuser (b. 1734), who emigrated from Fuerschbach-Hanau in the province of Alsace, arriving in Philadelphia October 26, 1767, on the ship *Britannia*. With them were their four sons: Johann ("John"), the eldest (b. 1758); Carl Otto (b. 1760); Johann Petrus ("Peter") (b. 1762); and Andreas ("Andrew") (b. 1764). A daughter, Maria Catherine, and two other sons, Johann Frederick II and Johann Georg ("George"), were born in America.

Unlike many of their countrymen, these Steigerwalts entered America as freemen, apparently able to pay their own way; but financial concerns forced them to indenture their seven year old son, Carl Otto, to a farmer in Greenwich Township, Berks County. Bound for fourteen years, at twenty-one he was released from service and, following custom, he received a suit of homespun, a pair of shoes, and stockings.

Indentured servitude was a common eighteenth century practice and can perhaps be thought of as limited slavery. Once bonded, the servant and his labor, legally and literally, belonged to the master for the duration. For adults, this was usually seven years; for a child, like Carl Otto, his fourteen years was common. Many German immigrants arrived as indentured servants, having sold their services to pay for their passage. While family tradition holds that Carl was "bound ... to earn some of the money his father had to pay to bring the family across the ocean," his indenture, no doubt, actually provided his parents with a stake to establish themselves in farming, first in Berks County.

As with typical German families, their early history is sketchy. The Revolutionary War brings them a bit more into focus. Peter served the Patriot cause, first in Captain Jacob Ladich's Company in Colonel Ely's Battalion of the Berks County Militia. He apparently later joined the 1st Battalion, Northampton County Militia in 1785 under the command of Lieutenant Colonel Michael Brobst.

Tradition suggests that Peter's brother, Carl, also served in the Revolution, but no records verify this.

The role of the Germans in the American Revolution is little known, in a positive sense. "German" and "Revolution" to many means "the Hessians," the mercenaries George III hired from the Landgrave of Hesse to fight against the Revolutionaries. Few realize that many Hessians captured in the colonies were bound out, as was the custom at the time, as laborers or craftsmen to local merchants and farmers. There were no Prisoner of War camps. In Pennsylvania, many Hessians ended up bound to German speakers, like themselves. A craftsman-Hessian, bound to the farming Kershner family of Lehigh County, did elaborate baroque plaster ceilings now in the Winterthur Museum. Many Hessian soldiers married into the families of their masters. At the end of the Revolution, a number of others preferred to remain in America where they could find better future prospects than in Germany. Many became farmers and more than a few of the Dutch, especially of eastern Pennsylvania, have some Hessian in their ancestry.

While the important Battle of Germantown was fought near Philadelphia, and that city itself was occupied by British troops, much of the Dutch heartland went peacefully on its way during the Revolution although, like Peter Steigerwalt, many German men and boys served in the Revolution. The graveyards of most early churches in the region are proud to contain graves of veterans of the Revolution, often proudly decorated with small American flags by local chapters of the Daughters of the American Revolution. Many of Peter's descendants are proud members.

As the years passed after the Revolution, the next generation of Steigerwalts continued establishing themselves as successful small farmers near the present town of Andreas in what was then Northampton County. The ancestral homes Peter, Carl, and Andrew built for their families, first log, were replaced by substantial stone residences and some are still owned by their lineal descendants. All three brothers were active in the Ben Salem Lutheran Church in Ashfield (near Andreas) where many of the family remain members today. Other descendants, like the author's in-laws, are buried or own burial plots in the Ben Salem graveyard.

Raising tourists is an important crop in Dutch Country. The Mennonite Eby family owns an eighth generation, 80-acre dairy farm near Paradise in Lancaster County. Since 1969 the family has operated Eby's Farm Bed and Breakfast, which is housed in two buildings: an 1814 brick farmhouse and 1999 Cape Cod. They have had guests from all over the world. *Lancaster-York Heritage Region*

"Wisteria on a Wall" was painted in 1871 by Germantown artist George C. Lambdin (1830-1896). The vine was named for Casper Wistar, M.D. (1761-1818) by botanist Thomas Nuttall. Wistar's ancestry traces to Germantown, where an ancient vine grows in the extraordinary Anglo-German garden of the family's ancestral home, "Grumblethorpe." *Schwarz Gallery Philadelphia*

The picture that emerges of the clan over the years is of people with large families who remained close to home. Most were farmers and/or craftsmen. Daughters married farmers or craftsmen. The records note weavers, blacksmiths, lumbermen, teamsters, storekeepers, tavern owners, teachers, and occasional ministers. Frederick, Carl's son and former indentured servant, stayed on the family homestead and married Maria Hettler. The couple produced twelve children, five sons and seven daughters – all of whom grew to maturity. Their valley must have been a healthy location and their genes sound because it is interesting to note how many children in these large families lived to maturity.

Frederick and Maria's son, Charles, was quite the entrepreneur. He became a teamster, as related in Pansy Steigerwalt's *The Steigerwalt Family 1767-1979:*

About that time he built a dam across the Lizard Creek south of his home. He built a saw mill, and the second son, Reuben, became the saw miller. He made his timber into boat luber [sic] which was hauled to Weissport. The lumber was hauled to the top of Mahoning Mountain by a team of four horses and taken the rest of the way by a team of two. A few years later he built a cider press; the builder was John Dieter Heintzelman, an expert woodworker.

People who wanted cider had to make it themselves at this press and were customarily allowed two days for it. They generally cooked apple-butter at the same time, using a 32-gallon copper kettle and a "patented" ladle, ad [sic] contrivance with four blades, which kept the apple-butter from sticking to the bottom of the kettle and burning. The cost for using the press was 6c for a barrel of cider and 25c for the use of the kettle.

The oldest son became a weaver, and as the younger sons grew up they became teamsters. The father saw to it that each one had something to do. He had a flock of sheep, using the wool for clothing and the mutton for food. He also raised flax for clothing, which the women made during the winter months.

Heidelberg, Germany, as it appeared in the eighteenth century. Modern Germany did not exist until 1870. Before that Germany consisted of hundreds of political entities. Many Pennsylvania German Church People trace their ancestry to the Palatinate, *die Pfalz,*whose principle port, Heidelberg, was a starting point for many an emigrant's journey. *The Cottage Collection*

An old world dream is remembered as a farm woman in typical dress pushes a wheelbarrow down a farm lane past an assortment of log buildings. Chickens roost on the rail fence to the left. Note how the growing tree has deflected the fence. *The Cottage Collection*

The next major event to bring the clan into historic focus was the Civil War. We learn that during the hostilities, Charles' eldest son, also named Charles, "was living in Tamaqua and was drafted into the service. Levi, the younger brother, was also drafted, but since he [already] had a family, his younger brother Stephen took his place." In the author's Steigerwalt kinship, the most vividly remembered of the Civil War veterans was Daniel Kressley (whose grandmother was a Steigerwalt), who survived the bloodbath that is remembered as the Battle of Antietam. Returning to his ancestral region, he farmed and married and had a large family, including a daughter, Esther Suzanna, who was born in the Lizard Creek Valley in 1885 and married local farmer and woodcutter, Franklin Steigerwalt. "Mama," as she became known to generations in the Valley, lived in her ancestral region until her death in 1983. She exemplified the basic conservatism of many of the Pennsylvania Dutch. She never had traveled more than fifty miles from her home and, although the railroad traversed her farm, she had never been on a train. All of her long life, English was clearly her second language. As she aged, and long before her death, she spoke English only when she absolutely had to do so.

Even from the earliest days of settlement, however, there were others who could not wait to leave the valley – some drawn by the prospect of adventure, others by the promise of cheaper land, and yet others by the quest for higher education. Indeed, tradition held that one of Johann Frederich's sons "migrated west" shortly after the family settled in the Lizard Creek Valley. Later genealogical work traced the adventurous soul to Indiana where his descendants still live. World Wars I and II also took many Steigerwalt men out of the valley, and while they saw the world, most returned home, as did many of the Steigerwalt men who had worked for the railroad companies of their day.

When you look through the various volumes of Steigerwalt genealogies, you are struck by the paucity of surnames before World War II. They reflect a society that was very traditional and devoutly Lutheran.

Typical of the transitional families within the Steigerwalt clan are the author's in-laws, distant cousins, Eugene Oscar Steigerwalt and Florence Caroline Steigerwalt, who were married in 1936. Both came from large families and both were educational pioneers in their families. Florence, who grew up on a traditional farm, was the first girl in her valley to attend high school and she went on to college, graduating from Kutztown Normal School (now Kutztown University) as a teacher. Eugene, the grandson of farmers (his parents were teachers), worked his way through Muhlenburg College and the Lutheran Theological Seminary in Philadelphia as a butcher. He was ordained in 1933. Interestingly, he commuted from Lehighton to Allentown and later to Philadelphia by trolley car – an inter-urban system destroyed by the automobile. As mentioned earlier, both came from large families and most of their siblings stayed in the region or returned back home. Eugene, like his father before him, always maintained large, and occasionally huge, vegetable gardens.

The Steigerwalts had five children, and as they came of college age, the minister and the teacher, facing overwhelming college expenses, moved to California where tuition was free and where Florence earned an M.S. in Library Science from USC. As retirement approached for the Reverend Steigerwalt, the family, with one high-school-aged child in tow, returned to Pennsylvania where Florence became Government Documents Librarian in Harrisburg and the Reverend Steigerwalt became pastor of two Lutheran churches – one in Port Royal and the other in the tiny crossroads town of Nook, both in Juniata County. Both congregations had a high proportion of farm families.

When the author joined the Steigerwalt family circle in 1970, attendance became mandatory at the 4th of July and Labor Day picnics at Lochner's Grove in Carbon County, where many of the hundreds of descendants of Ambrose and Alvesta Steigerwalt gathered. Most lived nearby; many traveled to be with their kin. Even today, Dutch country is dotted with these groves – some privately owned, others owned by towns.

Lochner's Grove, now closed, was located down a dirt road off Route 895 in Andreas. It was primitive. Toilets were outhouses, but in a nod to modernity, there was electricity. The grove had several pavilions. The largest housed the card games (mostly pinochle) and the major food consumption area. Another covered a grill, where the elders grilled frankfurters and hamburgers all day. The third housed barrels of beer for adults and root beer for children. Best of all, from the kids' point of view, there was also an old freezer stocked with tubs of ice cream and topped with boxes of cones and scoops.

There were amusements: horseshoes and quoits, and large multi-generational baseball games. This was a time to visit, to gossip, and to eat and eat. Cousins, aunts, and uncles as far as the eye could see. The elders' grill supplied snacks all day, not serious food. At about noon, real eating began, as each family carefully emptied the contents of their baskets and coolers onto the table. A veritable catalog of Dutch dishes emerged: sausage, ham, chicken, baked beans, three bean salad, coleslaw, potato salad, Chow-Chow, pickles (sweet and sour), pickled eggs, deviled eggs, and the homemade pastries – pies, cakes, and cookies. Many were made with homegrown ingredients.

Clan culture dictated that families sat together with their food in front of them and that picnickers were expected to walk along the tables and lean over and help themselves. In the 1980s, as the older generation died off, the picnics declined, and so did the quality of the food. More and more store-bought food arrived – supermarket platters of meats and cheese, cakes, and salads.

Since no one was about to compliment you on your supermarket purchase, meals became buffet style – the dishes often interchangeable and anonymous. In the early 1990s, Lochner's Grove closed and a *coup de grace* was given to a waning picnic. Other Steigerwalt picnics still go on. The Steigerwalt Family Reunion held its 95th edition on July 26, 2003, "Rain or Shine" at Zion's Grove. "Look for the Steigerwalt Reunion Sign." Today some Steigerwalts still farm, but even more have joined other segments of society, even while many continue to live in their home region.

While the Pennsylvania Germans are conservative, among the most successful of them, as suggested earlier, are progressive leaders within agriculture. In line with agriculture in general, the old-fashioned family farm where the family grew most or all of its own food and a whole range of crops has largely disappeared. Successful farmers of Dutch ancestry can be successful indeed. An exemplar is Luke Brubaker, who was voted "2005 Country Folks Keystone Farmer of the Year". Luke's father, Elam R., the son of a farmer who also worked as a day laborer and later as a part-time telephone repair man, scraped together enough money to buy a forty-acre farm near Mt. Joy in Lancaster and eight dairy cows in 1929, right after the stock market crash, which plunged America into the Great Depression. Elam and, eventually, his two sons Jim and Luke, literally grew the business. After the brothers acquired the farm, Luke bought out his brother, whose heart was not in modern agribusiness. Jim today runs a bed and breakfast on his own farm nearby. Luke clearly had his eye on the new world of super-sized production and since the 1960s his enterprise has grown dramatically. The Brubaker farm now covers 1500 acres devoted to cropland (mostly grains), dairy cattle, and poultry. The milk from his 600-cow herd is marketed through the Mount Joy Farmers Co-op and he raises 250,000 chickens for marketing giant, Tyson Foods. Luke's wife, Barbara, who wasn't raised on a farm, has taken to the family business and does the operation's secretarial work. Their two sons, Mike and Tony, both college graduates, help run the enterprise. Two of the ten grandchildren in the third generation show signs of wanting to following in the family business.

Committed to farming and farmland preservation, the Brubakers have preserved all of their 1500-acre farm through the Lancaster Farmland Preservation Trust. With suburbia pressing in on him from all sides, Luke still feels optimistic. "I think," Brubaker says, "that agriculture in Lancaster is here to stay." He is proud that, "We have some of the best soils in Pennsylvania," and that "There are a lot of good farmers doing a good job."

Public service-minded and cosmopolitan, Brubaker is a two-term member of the Pennsylvania Milk Marketing Board, was a board member of the Lancaster Farmland Trust, and is past chairman of the East Donegal Township board of supervisors. He has also traveled to Bolivia, Russia, and the Ukraine on missions to help local farmers increase their production.

His world, Brubaker notes, is changing rapidly. "… technology moves so fast and costs are going so high … [yet] consumers are expecting us to provide them with reasonably priced food …." The reasons he gives for staying in farming, set to music, could be the anthem of the Pennsylvania Farm. "Farming is one of the greatest things I can do with my hands and my body … Just to be out in the fresh air and see the crops growing and see my children and grandchildren grow up on a farm," he is rhapsodic. "It might not always be the most profitable to be a farmer, but it's a way of life that's hard to beat."

The principal gardens of the rulers of *die Pfalz* were formal and geometric. This illustration from *Hortus Palatinus* shows the gardens designed by Salomon de Caus, which had been created c.1613-1618. Order in gardens would become a hallmark of Pennsylvania German gardens. *The Cottage Collection*

Nostalgic postcards printed in Germany were very popular in Dutch country. Many have floral and/or garden or farm references. The birthday card shows a picturesque country village, while the lady at the well finds more than water in her bucket! The view of "The Meadows" was overprinted here in America with a glitter-laden glue "Greetings from Sellersville," a Bucks County town which had and retains a substantial Pennsylvania German population. *The Cottage Collection*

The Americanization of the Dutch is seen very clearly in this Lewis Miller watercolor of a visit to the Spangler Farm in rural York County by young Lewis (1795-1882) and his father, a music teacher in York City. The visitors are in fashionable attire; their hosts are dressed as old world peasants. The Spangler house is archaic as well; it is half-timbered (see Chapter IV). Note also the farm fences and the free-range chickens in the distance. *Historical Society of York County*

This half-timbered farm house was built several miles from Landis Valley by a Landis in the early eighteenth century. The center chimney house was clearly in decay and ancient looking when this photograph was taken c. 1890. *Landis Valley Museum Archives*

The Moravians are a church people whose women wore especially interesting garb. Many were highly educated for their day, often very knowledgeable about plant lore. Johanna Ingerheidt Schmick and Jacob Schmick were painted c.1754 in Bethlehem by Johann Valentine Haidt (1700-1780). Mrs. Schmick holds a distaff, an implement used for holding home-processed wool or flax in spinning, which is also a symbol of the female domain. Her blue ribbon denotes that she is married. Other ribbon colors would denote a widow, a betrothed woman, or an eligible virgin. *The Moravian Historical Society, Nazareth, Pennsylvania*

"Young Mr. Faul" by Jacob Maentel (1778-?) was painted in Lancaster County, c. 1835-1838. It is a reflection of the pride of a young Pennsylvania German who is obviously gratified with his fashionable dress and his fine horse. Lesser attention is given to his very Germanic log house and barn, beyond the fence. *Collection American Folk Art Museum, New York; promised gift of Ralph Esmerian; P1.2001.12; Photo © 2000 John Bigelow Taylor, New York*

The Hessian presence in Pennsylvania German culture is largely invisible, but a captured Hessian soldier, who in civilian life had been a plasterer, was bound to the Kershner family, prosperous farmers near Wernersville. While technically a prisoner of war he did the elaborate baroque plaster ceilings in his master's fine stone farmhouse. The rooms are now at the Winterthur Museum in Delaware. *The Henry Francis du Pont Winterthur Museum*

The Amish typically are tri-lingual. Most speak and/or understand English, German, and Pennsylvania Dutch. On occasion their language is English while their grammatical construction is Pennsylvania Dutch. *The Cottage Collection*

The Plain people are among the last tobacco farmers in the region. In this 1970s view, Amish and Mennonites harvest the crop. The woman in the foreground spears tobacco plants onto a lath. *The Cottage Collection*

Late nineteenth century hand-tinted photographs of Amish are very rare. An elderly Amish couple pose with a more worldly man holding a cat. Their farm is moderately well tended. In the left is a bank barn with a large haystack in the barnyard. Behind the woman is a pigpen, *der Seischtall* or *der Seiben*. The animal yard is to the left. Photograph was taken by Wilson Studios of Harrisburg and York. *The Cottage Collection*

An Amish woman walks down a driveway lined with ornamental stones and homemade planters filled with petunias (characteristically of one variety and color) and miscanthus grass (*Miscanthus Sinensis cv. "Gracillimus"*) popular in Dutch Country since Victorian times. Note that while her garb is traditional she wears very modern footwear. *The Pennsylvania Dutch Visitors Bureau*

Barn-raising is the most popular image of the Amish. It is a symbol of their community caring. When a barn burns or a barn is needed for a new farm, the whole community will raise one in a day. Before this miracle occurs, professionals have assembled the materials and joined many of the timbers that are to be raised into place. Men work on the barn; women keep the menfolk's thirsts slaked and their stomachs full. *The Cottage Collection*

The Amish mix with the "English" world at farmers' markets and farm-related sales like this farm equipment sale in Bird-in-Hand, Lancaster County. In a 1930s view a Mennonite woman and three Amish stand in Center Square in Lancaster. To the right is the Romanesque Revival Central Market designed by James H. Warner and built under the direction of John Berger and J. Adam Burger in 1889. A vintage view inside a market house features young Amish salesclerks. *The Cottage Collection*

A cross between the church people and the sect people, the Landis family, whose home would be preserved as part of the Landis Valley Museum, is seen in two late nineteenth century photographs taken outside their farmhouse. The cast of characters is father Henry Harrison Landis, mother Emma Diller Landis, older son Henry Kinzer Landis, younger son George Diller Landis, and the family darling, daughter Nettie May Landis. Henry H.'s family was Mennonite, but Emma Diller's were church people. The family all loved animals, which show up in many photographs. *Landis Valley Museum Archives*

Steigerwalt albums are filled with group pictures of large families taken by professional photographers. Most appear to date from c.1915. The Moses Steigerwalt family apparently posed against a barn wall. Another group, with young girls in starched ruffles, was photographed in front of their ancestral stone farmhouse. *Steigerwalt Family Collection*

The twelve grown children of Alvesta and Ambrose Steigerwalt pose on a family farm, c.1950. From left to right "Genevieve, Marguerite, Naomi, May, Fred, Paul, Stella, Viola, Anna, Tom, Andrew, Daisy, Laura." The dog's name is lost! *Steigerwalt Family Collection*

Daniel Kressley went off to the Civil War as a teenager. Surviving the Battle of Antietam, he returned home, farmed, married, and sired a large family. He and his wife are surrounded by their eight adult children in a group portrait photographed, c.1918. Esther Suzanna, "Mama", who was by then married to farmer and woodcutter, Frank Steigerwalt, kneels at the right. *Steigerwalt Family Collection*

The Ambrose Stahler Steigerwalts pose for an informal snapshot on the farm while eating watermelon, c. 1925. *Steigerwalt Family Collection*

Sunday afternoon was a time for visiting or just sitting on the front porch. Informally posed, this farm family's house has the two front doors typical of many in the Pennsylvania German Region. *The Cottage Collection*

Good food, and lots of it, is widely appreciated by the Dutch. Beginning in the 1930s, a Berks County innkeeper started to appeal to tourists with his all-you-can-eat, or *Alles os du esse Kanst*, family style dinners. The Shartlesville Hotel lasted until the 1990s. Haags Hotel, a little newer and down the street, still exists, serving traditional fare. Modern tourist promotions continue to capitalize on the local fare. *The Cottage Collection; Pennsylvania Dutch Visitors Bureau*

There were always some Pennsylvania Germans who bought their food. Farmers markets in towns date from the early eighteenth century. By early in the twentieth century, even small towns in rural regions offered a variety of dealers in foodstuffs. Advertisements from a flyer printed in Orwigsburg in Schuylkill County offer a wide range of foods, both fresh and canned. *The Cottage Collection*

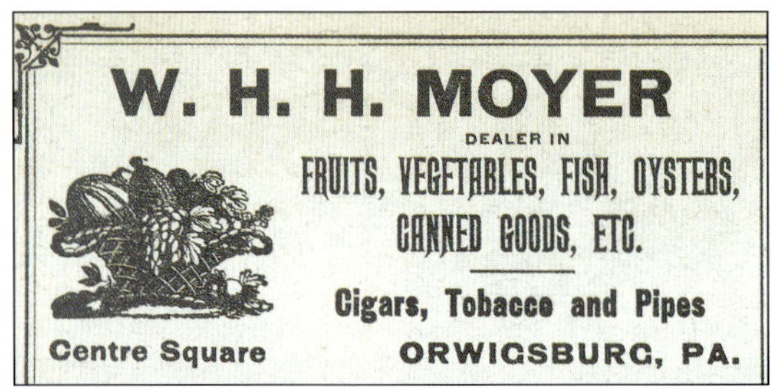

Adapting to new trends in agribusiness, Mount Joy dairy and poultry farmer Luke Brubaker farms 1500 acres with 600 cows and annually produces 250,000 broilers. He poses by a John Deere tractor. Note the huge steel silos behind him – the sure sign of a successful, large scale, farming operation. *Lancaster New Era, Marty Heisey, Photographer*

The Dutch are also owners of successful independent groceries. Some, like Oregon Dairy, are direct outgrowths of farms. Here storeowner Calvin Hurst (center) of Oregon Dairy is surrounded by Mike Musser (from left) and Greg Musser of Musser's, Jim Eshleman of John Herr's, Larry Darrenkamp and Dave Darrenkamp of Darrenkamp's, Ron Monahan of Foodland, and Willie Hurst of Oregon Dairy. All but Monahan have Dutch surnames. *Lancaster New Era, Richard Hertzler, Photographer*

Tradition, and reminders of tradition, can be seen everywhere in Dutch country. A decoration on a mill wall is a case in point. "This mural, painted by Artist Wayne Fetro, was done in April 2001 to celebrate our 76th Anniversary, and to announce to the community that premium quality **popcorn** is our primary business. The popcorn harvest scene features founder, Alvin J. Reist, driving a 1920s vintage Fordson tractor, which was the first tractor he used. His great-grandson proudly shows off the new crop. The **Hi-pop** trademark dates back to the 1930s and is still used today. David A. Reist, President". *Reist Popcorn Company*

The farm stand, where farmers can sell directly to the consumers, is an increasingly popular way to raise farm income while providing customers with very fresh produce. Wayside Produce near East Earl in Berks County sells vegetables and flowers in an August photograph. *Lancaster-York Heritage Region*

II. Developing the New Eden

Religion was a strong motivational factor in the period of settlement and many emigrants from the British Isles and the Continent viewed the undeveloped, largely unsettled, and unexplored North American continent to be nothing short of the "New Eden." Here was to be a chance, in the eyes of many of the pious and not a few of the impious, for mankind to start over again: to attain the perfection of an earthly paradise that had existed before the fallibility of Adam and Eve led to the expulsion. Here they would see the *Lilienzeit,* the Millennium of God's Rule. Folklorist Dennis Boyer relates that the pious Dutch worked towards this great end "… by making the journey and building … [the] new land. Every barn built, every stump pulled and every hope chest … painted for a daughter was a prayer made on behalf of that earthly paradise."

Alas, the vision for a fresh beginning was not to be completely fulfilled. Mankind, in general, was really more anxious to replicate familiar lifestyles across the ocean, than to repent and start anew in a state of grace. Even though America would be sullied, images of America as a new Eden would abound in art. Depictions of the Garden of Eden figure prominently in American, and especially Pennsylvania German, art and this imagery of a new and special place is particularly vivid in the paintings of Bucks County Quaker painter Edward Hicks (1780-1849), whose many versions of "The Peaceable Kingdom" and "Penn's Treaty" show America, and especially Pennsylvania, as a place blessed by the Creator. To the Pennsylvania Germans, primarily a rural and agricultural people, the new land was also greatly appreciated. In 1757 Gottlieb Mittelberger observed "… that Pennsylvania is the farmer's heaven …." If not heavenly, it was close to perfect. The Berks County town of New Jerusalem does not stand in an arid locale as does its old world counterpart, rather directly ahead of it are the lush agricultural lands of the East Penn Valley. Cultivated by generations of Mennonite farmers, the region would be the envy of any European farmer. In Lancaster County tourist-laden Route 30 passes through the town of Paradise itself, much of which in its modern incarnation, perhaps, looks more like a commercial Hades. This commercial strip is occasionally relieved by an especially picturesque church and the Classic-Revival-style, grandiloquently named, "Christ's Home for Children." However, the countryside away from the highway is a fertile, beautiful, pastoral vision. Elsewhere you can find a suburbanized Eden, complete with a motel and gas stations.

To the folk mind there was always a story to explain the shortcomings of the New Eden. A particularly rich one is a tale recounted by Dennis Boyer, which he learned from his Union County grandmother, a pow-wow doctor who practiced mystic healing.

Pennsylvania German decoration is often nature-based. Some might claim that it is a visual prayer made on behalf of a belief in an earthly paradise. Gloriously emblazoned with birds and flowers, this chest or *kist* boasts similar motifs to those often found on *fraktur* and upon many other Dutch artifacts. Dated 1801 and signed "John Seltzer", this chest was painted near Jonestown (once Lancaster, now Lebanon, County). *John and Mary Metzger; photograph Heritage Center of Lancaster County*

> She said there was a city by the name of Heidelberg. The city was near a big river. Along that river there was a cave. The cave was like a big butcher kettle that caught all sorts of bad things coming down the river. There had long been a curse on that land, so there was plenty of bad stuff floating around – everything from murdered travelers to half-burnt witches. And their ghosts collected in that cave.
>
> When the devil heard that some of the people were going to go to Pennsylvania, he took the mud in that cave and made a big ball of pain and filth. He gave it power to be anything it wanted to be—human, animal, plant or part of the Earth. It was sent along on those boats to Pennsylvania.
>
> But it didn't turn out the way the Devil hoped. There were things in that cave muck that he didn't count on. Like goodness and laughter.

So far the good appears to be outweighing the evil influences. But the final verdict is not in.

Among the earliest educated Europeans to appreciate the Pennsylvania countryside were the Rosicrucians, the "Knights of the Red Cross," a group of mystics and philosophers who established a colony along the Wissahickon Creek, a tributary of the Schuylkill River near Philadelphia, as early as 1694. By the early eighteenth century they had planted a garden of medicinal plants, which was an expression of their major belief that every plant is a portion of the "Master Mind," the Creator. The Rose-cross knights, also known as the "Hermits of the Wissahickon," claimed that through the proper use of their medicinal herbs they could "prolong life indefinitely." Primarily made up of seekers from the British Isles, they soon also attracted Germans as well. One of these was Johann Conrad Beissel (1732-1768), born in Everbach in the Palatinate. Young Beissel was outspoken in his radical religious views and this led to his banishment from the Palatinate. Fleeing Europe, like many of his compatriots, he settled first in Germantown. There he soon became associated with the Rosicrucians, among whom he lived for some time, perhaps as a member. Always charismatic and a poor follower, Beissel soon retreated to live the solitary life of a hermit in Muehlbach, in present Lebanon County, but disciples insisted on following him. His magnetic personality and the strict seventh-day baptism which he espoused, became the basis for "Economy" at Ephrata in Lancaster County, a religious communal society he caused to be founded in 1732. The "Ephrata Cloister," as it has come to be known, consisted of three orders: celibate Sisters, celibate Brothers, and married householders, all of whose labors helped support the cloistered orders. Under Beissel and his disciple Peter Miller the order grew to encompass several hundred largely self-sufficient brethren.

Combining imagery drawn from nature, superstition, and commerce, the Pennsylvania Dutch hex signs have spawned a minor industry. They can be found on everything from restaurant menus to rolls of toilet paper. On the obverse of this 1970s card: "Prof. Johnny Ott, Hexologist from Lenhartsville, Penna. And Jacob Zook, The Hex Man of Paradise, Penna., bring you "Good Luck Signs" for every situation from "sore feet" to the "farmer with unhappy pigs" to "Mother-in-law troubles". *The Cottage Collection*

Agriculture was the basis of their prosperity and the order erected its own gristmill. Members also grew flax and raised sheep to provide their simple clothing. Aside from occasional spiritual "love feasts" where meat (usually lamb) was served, Beissel's followers were mostly vegetarians and so had extensive vegetable gardens. Agriculture, then, provided the basis of support for the distinctive music, calligraphy, and printing for which the community was to be renowned. For a brief time the Cloister was home to Conrad Weiser (1696-1760), the famed Indian agent whose own fortune was based, in part, on the extensive well-run farms he owned in Berks County. The community's decline began with Beissel's death on July 6, 1768, but small numbers of his followers at both Ephrata and its offshoot, Snowhill Monastery in Franklin County, maintained their distinctive lifestyle well into the nineteenth century. Ephrata's surviving buildings are the most significant Germanic Gothic structures in America. Maintained today by the Pennsylvania Historical and Museum Commission, the Ephrata Cloister has several recreated gardens as well as token plantings of flax and grains to give the modern visitor a glimpse of its agricultural past.

The hand-colored frontispiece of Gerard's *Herball*, 1597, is resplendent with mythical gardeners. The vignette of a garden and castle at the bottom shows the strong influence of the continent on English art and gardening. Gerard's *Herball* was well known and influential to the continental learned community. *The Cottage Collection*

Despite Beissel's apparent interest in herbalism, probably acquired during his residence with the Rosicrucians, agriculture and gardening by the residents of Cloister was directed mostly to survival, although they no doubt grew medicinals and flowers, some of whose forms are used in the decoration of their famed illuminated manuscripts, frakturs. More worldly, richer, and better educated were the Moravians whose settlements, especially those in Bethlehem, Pennsylvania, and Salem (now Winston-Salem), North Carolina, were famed for their ornamental gardens and their botanists as well as for their agricultural prowess.

The Moravians are a Protestant denomination founded in the eighteenth century, but one which traces its roots to the *Unitas Fratrum* (Unity of Brethren) Hussite movement in fifteenth century Bohemia and Moravia.

That the proper plants and gardens can create an earthly paradise is the theme of the frontispiece of John Parkinson's 1629 *Paradisi in Sole, Paradisus Terrestris* (Paradise in Heaven, Paradise on Earth). The work was widely admired and translated into various continental languages as well. *The Cottage Collection*

Surviving as an underground movement, "the hidden seed" was brought into public notice through the efforts of Count Nikolaus Ludwig von Zinzindorf (1700-1760), when a group of families following the traditions of the Bohemian Brethren fled Moravia in 1722 and settled on the Count's estate in Saxony. On von Zinzindorf's land, they founded the Village of Herrnhut, which would become the mother community of this group, and one that attracted Pietists from all over Germany. Central to this group was a lasting evangelical zeal and soon after the move to Herrnhut the first foreign missionaries were sent out to minister to Black slaves in the West Indies. Within twenty years missions were in places as varied as South Africa, Greenland, Surinam, and North America.

The pioneer Moravians in the thirteen original colonies began their work in Georgia in 1735 where Zinzindorf saw an opportunity to evangelize the Indians and to create a haven. Unsuccessful there, the group turned to Pennsylvania in 1740 where they quickly founded Bethlehem and Nazareth. A few years later, Lititz followed. The Moravians also relished the prospect of uniting Pennsylvania's German settlers – Lutheran, Reformed, or sectarian – in a union church. The effort ultimately failed, but during von Zinzindorf's fourteen-month American stay (1741-1743) he did succeed in establishing the Moravian church in the New World.

Moravians from Pennsylvania founded Salem, North Carolina, in 1753. Moravian communities, whether in Pennsylvania, North Carolina, or elsewhere, were carefully planned and developed. They often featured ornamental communal gardens open to the community and widely praised by visitors. This story has been detailed in James Murtagh's *Moravian Architecture and Town Planning in America*. The Moravians began as communitarians, with their society divided into "choirs," or groups, by age, marital status, and sex, where all lands and industries were owned by the society. This system would dissolve and privatization was the rule by the end of the eighteenth century. However, until the mid-nineteenth century, the successful Moravian communities were closed to all but Moravian residents. Today, the historic portions of their major American centers function as combination church-owned facilities, schools, and museums.

Edward Hicks (1780-1849), a Quaker, captured the Edenic image that many British Islanders and Germans envisioned for Pennsylvania. He celebrated this in the many "Peaceable Kingdoms" he painted – where animals and children co-exist in tranquility and in his visions of William Penn treating honorably with the Indians under the Treaty Elm. *The State Museum of Pennsylvania, Gift of Vivian O. and Meyer P. Potomkin, Pennsylvania Historical and Museum Commission*

The love of nature and a sense of whimsy is often seen in Pennsylvania German folk art. One of the best examples is seen in a watercolor by an Schwenkfelder amateur. A stylized rose bush bears three blossoms, one on each of its three branches, but it also bears a tulip. Fortuitously a mythic *distelfink* (goldfinch), or perhaps a parrot, perches on this botanic impossibility. *Schwenkfelder Library and Heritage Center*

Among the early Moravian botanists who had a great interest in native flora was Samuel Kramsch (1756-1824), who served for years at the Moravians' Nazareth Hall School in Nazareth, Pennsylvania. There he taught natural science to several important future Moravian botanists, including Christian Fredrick Denke (1775-1838) and Louis David von Schweinitz (1780-1834) before moving to North Carolina.

Not long after Kramsch arrived in Salem, he wrote to Humphrey Marshall (1722-1801), a botanical mentor in Pennsylvania. "I botanized hereabouts," he noted, "as much as time would permit it, and found a great variety of plants between here and my former place [Nazareth], although much more difficulties concerning the heat, and especially the insects called Tiks [*sic*]." Clearly he had left the true Eden. Among the plants Kramsch found was the Carolina allspice (*Calycanthus Floridus*), which he described as exuding the fragrance of strawberries and growing very well in low-lying damp areas, "... especially near the mill." The allspice bush, with its sweet smelling almost mahogany blossoms, became a favorite of the Pennsylvania German when it was introduced north, where it was often called "Sweet Smelling Shrub." Young girls liked to tie a blossom in a handkerchief to use as a sachet.

The most important of the Moravian botanists was Louis David von Schweinitz, the Bethlehem-born grandson of Count von Zinzendorf. Christened "Ludwig," he anglicized his name early and on his scientific work he often signed his name with the less aristocratic "de" instead of "von." A missionary who held important posts in both Bethlehem and Salem, he botanized extensively for many years. His specialty was the study of fungi and his *Synopsis Fungorum Carolinae Superioris* is the first important work published on American fungi and it has earned him the sobriquet, "Father of American Mycology." Schweinitz's singular scientific work, *Synopsis Fungorum in America Boreali*, which described 3098 species, was published in the *Transactions of the American Philosophical Society* in 1832. Over his life he published almost 4500 species of fungi, of which over 1500 were believed to be new to science. From his work he was also able to establish ten genera. Additionally, he was interested in flowering plants and was especially attracted to the genera *Viola*, which includes violets and pansies. One of his best scientific papers is on the genus Viola, which in systemic botany is a very difficult genus because of its complexity. The research was published in the *American Journal of Science and Arts* in 1822. As a result of his interests he collected a large botanical library and he left a 23,000 species herbarium, now housed at the Pennsylvania Academy of the Natural Sciences in Philadelphia.

The Garden of Paradise is a familiar subject of German and other continental artists of the Renaissance. This walled garden ablaze with flowers, abundant with fruit and water was painted by the Upper Rhenish Master in the fifteenth century. It is an idealized world of peace and contentment. Could this state be reproduced in the new Eden? The American version, drawn from a Pennsylvania German manuscript, is devoid of people but it is also protected behind a wall and is overseen by two peacocks, symbols of plenty. *Image, Private Collection; Free Library of Philadelphia*

Von Schweinitz and the other Moravian botanists all corresponded with the extraordinary Gotthilf Henry Ernest Mühlenberg (1753-1815), a member of one of the most accomplished and important Pennsylvania German families. Gotthilf was the son of Henry Melchior Mühlenberg (1711-1787), patriarch of the Lutheran Church in America, grandson of the Penns' Indian agent Conrad Weiser, and brother of John Peter Gabriel Mühlenberg, "The Fighting Parson" of Revolutionary War renown. Born in Trappe in Montgomery County, Gotthilf was sent to Germany to be educated. He returned to America in 1769 and the following year, at age seventeen, he was ordained. Serving as a pastor in Philadelphia when the British occupied the city Gotthilf, a patriot, was forced to flee the city, finding sanctuary at his birthplace in Trappe. Here in 1778, during his enforced sojourn in the countryside, he began the study of botany, which became his major avocation. In 1779 he was called to Trinity Lutheran Church in Lancaster where he would remain until his death thirty-five years later. He would also become the first president of Franklin College (now Franklin and Marshall) in 1787. Whenever he could, he botanized. In 1787, Johann David Schöpf (1752-1800), a German born and university-trained botanist, visited him and gave young Mühlenberg "... confidence in the soundness and value of his work." Through Schöpf's introductions, Mühlenberg would exchange letters and specimens with scientists in Sweden, Germany, Switzerland, England, and France, as well as maintaining his widespread American correspondence.

Mühlenberg's botanizing was only pursued when his professional duties allowed. Accordingly, his area of exploration was extremely limited. He did most of his fieldwork within a three-mile radius of Lancaster City, and by 1791 he had listed 1100 plants growing within this restrictive perimeter. He also kept a detailed calendar of flowering plants in his area. Maintaining scrupulous regard for correct nomenclature, he recognized the need for collaboration among scientists to compile a comprehensive flora of North America. Within his little sphere of direct observations he discovered and described about 100 species and varieties. Additionally, he gave great attention to the economic and medicinal values of the plants he studied. His *Florae Lancastriensis* (Flowers of Lancaster) was published by the American Philosophical Society in 1788.

Many of the early appreciators of American plants and cultivars of Old World materials were especially interested in medicinals as was Gotthilf Mühlenberg's father, the almost mythical Henry Melchior Mühlenberg. Before the elder Mühlenberg there had been Francis Daniel Pastorius, a founder of Germantown who "... in all probability [was] at least an armchair Gardener in the Old World," and who brought some horticultural books with him when he emigrated, along with seeds and grape cuttings. Among Pastorius's avocations were beekeeping and gardening, which included the cultivation of plants for medicinal usage. His gardening, as befits the rare university graduate in the colonies, had an intellectual bent piqued by an inquiring mind and the desire to experiment. While many have sung the praises of Pastorius as an agriculturist, he himself was more modest and wrote about merely creating a, "pretty little garden in which he grew ... chiefly cordial stomachic and culinary herbs." As a scholar Pastorius assembled the first herbarium in America in 1707 and left medical manuscripts, including one entitled "*Artznei buch*," a pharmacopoeia complete with notes on diseases and their remedies. He especially emphasized herbs native to Pennsylvania.

A later herbalist in Germantown was Georg Bensel, a shopkeeper by vocation but an avid collector of botanical lore, who apparently also authored a manuscript pharmacopoeia as Pastorius did before him. From a historical vantage, perhaps the most important eighteenth century herbalist was Christopher Sauer (1721-1784), also an apothecary and printer who printed the first American herbal "... yet," observes historian William Woys Weaver, who has published an annotated version of the book as *Sauer's Herbal Cures*,

> ... it was old-fashioned and scientifically out of date even as it appeared in print between 1762 and 1778. Issued under the title *The Compendious Herbal*, it was also one of the first American books to be published in serialized form over this sixteen-year period, but few complete copies of the bound herbal now survive. As a system of botanical medicine—cures and remedies based on the use of herbs—the Sauer herbal is rooted in the medical thinking of the late Middle Ages and Renaissance, yet in its own unique way it forms a bridge between that era and eighteenth-century colonial America. It is not a product of the leading thinkers of that period but rather a barometer of the mindset of the working class. It is a people's herbal, and that is what makes it so fascinating to read over and over again.
>
> Whatever its scientific shortcomings, this people's herbal is married to the land. Its very greenness—or better put, its pervasively green, medicine – provides a link to the woods, meadows, and kitchen gardens of the past.

Close to the hearts of the scientific and, often the practical minded, is the botanical garden. According to U. P. Hedrick, writing in his pioneering *A History of Horticulture in America to 1860:*

> The first botanic garden worthy of the name was established and maintained for some years by Christopher Witt in Germantown, Pennsylvania, in 1708. Witt (1675-1765) was an Englishman who came to

America in 1704 and joined the Rosicrucians in their colony on the Wissahickon. He was a physician, botanist, and a man of note in Philadelphia. Besides the legitimate professions of medicine and botany, he was a conjurer, fortuneteller, astronomer, and a professor of the black art of hexing. However, it is his botanic garden with which we are concerned, one that antedated John Bartram's famous garden, generally thought to be the first, by 20 years.

In 1718, Witt owned 125 acres of land, but how much was covered by the botanic garden does not appear. At first the plants grown were those used in medicine, but later he collected and grew plants to send to England, especially to Peter Collinson, the English Quakers botanist, by whom, as appears in his letters, Witt was held in high esteem. He was also a friend of John Bartram (1699-1777), with whom he had in later years much intercourse. Bartram wrote a long letter to Peter Collinson, giving an account of Witt as a man and botanist, under date of 11 June 1743. Witt, who died at ninety, was the last of the early Rosicrucians.

One of Witt's colleagues was a lesser known botanist Johann Kelpius, whom some sources have claimed had founded an even earlier botanical garden, but both Kelpius and Witt unquestionably collected many plants, some for pleasure and others for sale. Their collections were continued by Germans Christian Lehman and Melchior Meng. Meng, the collector, is thought to have first grown the *Magnolia macrophylla,* or big-leaved magnolia, in a Pennsylvania garden. Lehman, who did many things, was at once a surveyor and a translator, "... of English and High Dutch writings with Exactness ... in either of the said languages ... was to found a nursery in Germantown which is one of the first in the English Colonies."

Lehman advertised in Philadelphia newspapers, the *Pennsylvania Gazette* and the *Pennsylvania Chronicle,* giving descriptions of his stock and approximate times for their plantings. In 1768, for example, he was offering for sale a "Choice parcel of well-grown young English walnut trees as well as Pear and Apricot and a curious variety of the best and largest sorts from England of grafted plumb trees ... as well as a great variety of beautiful double Hyacinth roots and Tulip roots." Lehman also advertised being willing to "... on request and if bespoke in time" to provide custom lots of "... curious plants, shrubs, and seeds of the growth of this climate in such a manner as best secures them according to what country or climate they are designed to be transported."

Printer unknown. Adam and Eve, c. 1800. Adam and Eve were popular fraktur subjects, especially for German-language printers of southeast Pennsylvania. Adam's obvious interest in Eve's attributes would have appealed to the Pennsylvania German sense of humor, which is often described as "earthy." *Private collection.*

Images of Adam and Eve in the Garden of Eden, in both hand-colored graphics and watercolors, are very common in nineteenth century Pennsylvania German folk art. One is even embellished with the tulips found on so many pieces of pottery and furniture as well as *fraktur*. Unlike the occupants of the Garden of Eden, the Pennsylvania Germans learned their lesson – and they stayed on in Dutch Country. *Private Collections*

He was hoping, in a twenty-first century phrase, "to go international."

The saga of English plant explorers is well known; that of the Germans is less familiar and often revolves, in some way, around Pennsylvania. At the turn of the nineteenth century one Matthias Kin, who was described in the *Pennsylvania Gazette* as "... an eccentric person who collects seeds," was sent to America by a group of German plant lovers to collect promising garden plants. The local financial agent for the Germans was Melchior Meng, to whom Kin gave many plants, which Meng planted around his Germantown home. These grounds, almost twelve acres in all, were to have an interesting history. After Meng's death, the property passed to the Wistar family and then it became the home of Thomas Meehan (1826-1901), the great English born plantsman. Later it was to become Vernon Park, a Philadelphia green space. It was here that the *Magnolia macrophylla* planted by Meng grew and thrived into the middle of the twentieth century. Kin was a contemporary of the French born plant explorer Andre Michaux (1746-1802) and may have collected with him. Michaux introduced many European plants to Pennsylvania, including the Lombardy poplar, which would become a favorite tree on Pennsylvania German farms. Planted along lanes, its neat upright growth appealed to the German sense of order. Kin also collected many potentially valuable plants for the Landreth nursery in Philadelphia, founded by David Landreth, a Quaker, in 1784.

More important than Kim was Frederick Trougott Pursh (1774-1820) to whom, author U. P. Hedrick claims that, "... all American gardeners are indebted." German born, as were many of his compatriots, Pursh was a botanist, an explorer, a horticulturist, and an author. It was he who first brought the beautiful azalea *Rhododendron arborescens* into the world of gardening. In 1802 he took charge of the gardens at the Woodlands, the Philadelphia estate of William Hamilton, widely considered among the most beautiful in America. He left to continue his plant collecting – trips which would take him throughout much of what is now the continental United States. He would, for example, name the Pacific Coast Madrona *Arbutus Menziesii*. In 1807 he worked at the Elgin Botanical Gardens in New York and then in 1812 he went to England and studied herbariums and living collections of American plants that had been assembled there. In 1814 he published *Flora Americae Septentrionalis*, which drew on his experiences, and those of others, who had also been botanists on the Lewis and Clark expedition. His only other published work was printed long after his death, *A Journal of a Botanical Excursion in the Northeastern Parts of Pennsylvania and New York* (1869). Always peripatetic, he died in Montreal in 1820 where he had settled in preparation for an exploration of the flora and fauna of Canada.

Central to the appreciation of the flora and fauna of America, by Anglos and Germans, was the work of John Bartram (1699-1777) and his son, William (1739-1823).

"Arbeite und hoffe", "Work and Hope" is the motto of the Ephrata Cloister. It is seen here in a seventeenth century German graphic reproduced from Julius Friedrich's pioneering *The German Sectarians of Pennsylvania ... History of the Ephrata Cloister ...* (1899). Note the spade shape, still favored by many Germans. *Image, The Cottage Collection*

The Ephrata Cloister was founded by Johann Conrad Beissel (1732-1768), a religious mystic who had lived among, and might even have belonged to, the Rosicrucians, perhaps America's earliest herbalists. Essentially vegetarians, the Ephrata community had extensive gardens. Its buildings, medieval German in style, were constructed using the proceeds of its farms, mills, and printing press. *The Cottage Collection*

A Quaker born in Delaware County, John Bartram in 1728 purchased land at Kingsessing, about three miles from Philadelphia (now within the city limits) on the banks of the Schuylkill River. He began to farm this land and here he was to found his celebrated botanical garden which, if not the first in North America, was certainly to be the most influential. Bartram is celebrated for his correspondence with English and European gardeners and with scholars, notably Peter Collinson, an English Quaker. He became so well known in Europe that Linnaeus, himself, called Bartram, "... the greatest natural botanist in the world." Included among his correspondents were a number of German-American botanists. A companion on one of his many botanical forays was the seeming omnipresent Conrad Weiser, who traveled with Bartram as he explored central Pennsylvania and portions of New York. Bartram's journal of the trip was published as *Observations on the Inhabitants, Climate, Soil, etc. ...made by John Bartram on his travels from Pennsylvania to ... Lake Ontario in 1751*. Bartram preferred to travel in the autumn when he could most readily collect seeds and roots and young transplants that were most likely to survive. While he was to be the discoverer of many useful and ornamental plants, his most famous discovery was made during a 1765 trip through the South when he found the Franklinia (*Franklinia altamaha)* growing near the Altamaha River in Georgia.

Gotthilf Henry Ernest Muhlenberg (1753-1815) was one of the leading Pennsylvania German botanists. The longtime pastor of Trinity Lutheran Church in Lancaster, he mostly botanized within a three-mile radius from that community. Nevertheless, he did important botanical work and corresponded with other important American, as well as European, botanists. His portrait was painted by Pennsylvania German artist Jacob Eichholtz (1776-1842). *Phillips Museum of Art, Franklin and Marshall College*

John Bartram (1699-1777) was the central focus and informal clearinghouse for botanical studies in Colonial Pennsylvania among the English and the Germans. His famed botanic gardens begun in 1728 were visited by every early botanist of note, as well as by major political figures including George Washington and Benjamin Franklin. This eighteenth century sketch shows his house and a plan of his garden, which was a combination of botanical garden and nursery. No portrait exists of Bartram, who was a modest Quaker. *John Bartram Museum and Gardens*

A beautiful late-blooming shrubby tree with saucer shaped pure white flowers and large yellow stamens, the shrub's leaves turn a rich bronze in autumn. He brought the plant back to his garden on the Schuylkill where it thrived and in 1775 his son William, who would botanize with his father and would inherit the gardens, sent the tree to England. It is from the specimens at Bartram Gardens and the one sent to England that all the known plants in cultivation today are descended. No one has seen the species in the wild since 1790. Most famously, in 1765 Bartram was appointed to the post of Botanist to the King (George III) with a royal stipend of sixty pounds per year.

On an intellectual and aesthetic plane, the best aspects of Pennsylvania German and Quaker gardening and botany were brought together at the remarkable estate called "Wyck" in Germantown. The property, originally covering sixty acres, is today just 2 1/2 acres of gardens and landscaped grounds. Hans Millan (d.1698), a German Swiss Quaker, bought the first part of the property in 1689 and the farm-into-estate would remain in the family until 1973 when it was opened as a museum. Part of the house Millan and his wife built is incorporated into the great house that is now called Wyck, which was altered into its present form by the great Philadelphia architect William Strickland (1788-1854) in 1824. Very little is known about the grounds in the early period, but it is fair to assume that there were gardens in which to grow vegetables for the family and some herbs for medicine and flavoring. It is also probable that fruit trees were also grown because quite early Germantown was known for the quality and variety of its fruits. There were perhaps a few flowers sprinkled here and there, but most of the property was devoted to raising the farm crops of the era.

John Bartram's house has been preserved as a botanical shrine since the nineteenth century: first privately and now by the City of Philadelphia with the active help of a not-for-profit group, The John Bartram Associates. The early twentieth century view shows the house romantically smothered by vines. The later photograph shows the house, more clearly, in its original intent. Bartram himself did the stone masonry, complete with inscription on the plaques below the windows. There is strong historical evidence for the fig trees growing against the south-facing wall. A garden of American natives grows in the foreground. *The Cottage Collection; Author's photograph*

Documented information about the garden came into focus beginning in the 1790s during the residency of Caspar Wistar Haines (1762-1801) and his wife Hannah Marshall Haines (1765-1828). The full botanic importance of the place comes with the ownership period of their son Reuben (1786-1831) and his bride Jane Bowne (1790-1843) whom he married in 1812. Jane was a Quaker from New York whose family had also been very interested in gardening.

In January 1797, Hannah wrote a very interesting letter to her young son Reuben, then at Westtown School, a Quaker boarding school in Chester County, in which she chats about forcing flowers as well as vegetables:

> Peter desires I will tell thee he has caught two dozen of mice, and he says he takes great care of thy flowers. The daisey's look beautiful and the Hyacinths are blowing—Thy Father had two hot beds made yesterday for Cabbage. He has one box full up, and another saw'd a day or two ago. [Original spelling and punctuation.]

A month later Hannah could report that the cabbages were flourishing and, surprisingly, that, "... our peas are stuck planted and some of the beans," which suggests that a cold frame was being used. Hannah is also happy to report that one of their "... wall flowers [*Cheiranthus cheiri*] is in bloom." A close relative of the cabbage, the Wallflower was perhaps growing close to a stone wall that enabled it to bloom so early. The Germans, according to Christopher Saur, called it "*Gelbe Nelken-Viole*" and it was prized among colonists as both an ornamental and for various medicinal uses, especially by midwives. In this letter Hannah also lamented that new flower seeds were "too expensive to buy."

In her March letter to Reuben, his mother provides especially interesting information when she reports that, "Josey has begun to put fresh tan in the garden walks and today planted out some nice sallad plants." The use of tanbark on "garden walks" suggests that there were regular defining paths, which probably means that there was a geometric arrangement of beds of flowers and vegetables in a typical eighteenth century arrangement. The use of tanbark also reflects the fact that Germantown was a tanning community and, indeed, there were tanneries within smelling distance of Wyck. "Sallad plants" refers to greens either fresh or, more likely, cooked.

The Franklinia (*Franklinia altamaha*) is Bartram's most famous plant discovery and it was widely planted in Pennsylvania, including in Dutch country. Often fickle, there is a fine old specimen on the grounds of German-founded Franklin and Marshall College in Lancaster. The shrubby tree was photographed at Hope Lodge, an eighteenth century house in Montgomery County. The close-up is of a shrub in the author's Lancaster County garden. *Author's photographs*

Bethlehem in 1754 was engraved by English artist Paul Sandby (1725-1809). Note the extensive garden areas in front of the major Moravian community buildings. In addition to their ornamental value, the gardens "...were laid out with the provision that a portion was devoted to the growing of medicinal herbs for the [community's] laboratory and pharmacy... special attention ... [was] given to the study of flora of the region and men were appointed to collect herbs then known to have curative properties. These included various shrubs and plants, such as mandrake or May apple, snakeroot, sassafras bark, and berries." *Moravian Archives, Bethlehem.*

Caspar Wistar Haines had inherited Wyck from his mother, Margaret Wistar Haines, who died in Philadelphia during the devastating Yellow Fever Epidemic of 1793. Caspar then moved his own young family out to the Germantown countryside where conditions were healthier, as did George and Martha Washington who rented the Franks (Deshler-Morris) House nearby. Young Haines immediately began to improve his family property. He built a brew-house in 1794 and a barn in 1796 and he improved existing or, perhaps, created new gardens. As we have seen, he had both hot beds and cold frames and we know that he exchanged cuttings with his relatives and friends in Germantown. He would also buy about 150 new fruit trees from the pioneering William Prince's nursery in Flushing, New York.

On April 1, 1797, Hannah writes about the trees and more in one of her regular letters to her son.

> ...our flowers are all up in the garden, and the tree's are almost ready to Bloom, we have had a fence put up all round, and an arbour for the Honeysuckle to run on, and made it much larger than it was before, so that thee will see many alterraisons when thee visit's us. a fifth day thy Father received from New York one hundred fifty trees—Apples-Peaches-Pears and Plums, so that if we should live a few years we shall have plenty of Fruit—I expect also a large crop of rasberry's but hope by the time our fruit is ripe, we shall have thee to help us eat it...

Hannah's garden description is vivid and clear. It is an admixture of the practical and the ornamental—especially the "arbour for the Honeysuckle" which, following the custom of the time, probably had a seat under it where one could read and enjoy the pleasures of the colorful and productive garden.

When Caspar died in 1801 his son Reuben, then fifteen, was apprenticed to his uncle Christopher Marshall, a Philadelphia merchant. Reuben did not enjoy mercantile pursuits. When he turned "... twenty-one, his father's legacy in hand ..." historian Sandra MacKenzie Lloyd wrote:

> Reuben forever left full-time commercial enterprises and dedicated himself instead to "the pursuit of knowledge and the society of genuine friends." Rather than relaxing with the good fortune of his wealth, Reuben devoted his life to self-education particularly in areas of scientific study; to agriculture; to active participation in the Academy of Natural Sciences, the Franklin Institute, the Pennsylvania Horticultural Society, and the Philadelphia Society for Promoting Agriculture; to public service, notably in the implementation of school reforms; to the financial sponsorship of struggling artists and philosophers, including Rembrandt Peale and Bronson Alcott; and to the creation of a healthy, happy and stimulating home for his family.

This connected house-barn, *hofhaus*, is believed to have been the first home built in Bethlehem. Of this long gone building, Moravian leader Nikolaus Ludwig von Zinzindorf (1700-1760) wrote that he went into the next room "to tell the beasts of the Birth of Our Lord." This suggests that it was a house-barn. Census records reveal that such structures continued to be occupied in some Pennsylvania German areas until the late nineteenth century. Note the open bulkhead door to the right, which led to a storage cellar. A rain barrel collected water. To the right a picket fence encloses a garden. The gate abuts the house. *The Cottage Collection*

The Third Store in Bethlehem was erected in 1794 and subsequently enlarged. It was demolished in the 1920s. Note the enclosed garden and drying yard to the left. Behind the picket fence an arbor is clearly visible, as are supports against the house for grapes. The undated photograph is nineteenth century. *Moravian Archives, Bethlehem*

Reuben's wife, Jane Bowne, was the perfect complement to the country gentleman-scholar who Reuben was becoming. She is described as being, "Well educated and equipped with a quick wit." In addition to being a loving wife and devoted mother she was also creative and her "... passion for gardening complemented her husband's interest in agriculture and natural history."

Two years after their marriage in 1812, the young couple who had first lived in Philadelphia started spending their summers at "Wyck," which Reuben had renamed in 1807. The setting was still primarily rural, but the grounds had deteriorated during Hannah's widowhood because they had been rented to tenant farmers for over ten years. In June 1814, Jane wrote to her sister Hannah Collins about summering at Wyck.

> ...I find our situation here much pleasanter than I anticipated ... our outdoor kitchen answers very well. We have not however many beautiful views or a fine garden to show our visitors, as the house is immediately on the street—the gable end of it—and the farm extends behind it—the position of the ground does not admit of much ornament—and that which was once bestowed on it is now almost obliterated by a succession of neglectful tenants—we have plenty of excellent vegetables and there will be plumbs and apples but no peaches—a very pleasant shady yard in front where my little Sarah plays quite half her time and where she is now sitting in a wheelbarrow having just had a ride (here let me acquaint thee that the yard is made private by a stone wall next the street lest thee shd be shocked that we do not more maintain the appearance of citizens) Whilst I am seated in a most delightful hall where we live almost entirely it is 20 feet wide with two huge barn-like doors open to the yard in front and garden back—it is always cool here for there is no glass excepting a single row of panes over each door ...

"Schnitz House, so-called because it was here that Moravians of olden times prepared their winter apples for drying and storage," reads the obverse of this circa 1900 view of a still existing Bethlehem landmark. Long used by Moravians as a dwelling house, it retains a picket-bordered garden to the right, but the lawn features a circular flower bed, often seen on German lawns after c. 1850. One such ornament is mentioned in *Tillie: A Mennonite Maid* as quoted in Chapter IV where we are told that, "A grass-pot ornamented with a circular flower-bed ..." decorated Tillie's family's farmstead. *The Cottage Collection*

Nazareth was an established Moravian town by 1757. In the left foreground is the community's large vegetable garden, *die Lott*. The log building was a *Familienhaus*. Next is the still existent *Gemeinhaus*. Behind it is a small log building, probably a smoke house. The very Germanic looking Old Barn was built in 1775. The sketch was made by Lititz-born Moravian artist Rufus Grider (1817-1900) in October 1854. *Moravian Archives, Bethlehem*

A Moravian woman in garb was photographed in the nineteenth century behind the 1752 extension to the Single Sisters' House in Bethlehem. Note the flower beds built up against the buildings in a Victorian, rather than a traditional, configuration. The vines growing on the walls are probably Virginia Creeper (*Parthenocissus quinquefolia*). *Moravian Archives, Bethlehem*

John Lewis Krimmel (1786-1821) painted "John Gottlieb Heckewelder." Heckewelder, a Moravian naturalist, ethnologist, missionary, and author, who died in 1823, was in his late 70s when he was painted. He is leaning on a copy of his recently published work, *Account of the History, Manners, and Customs of the Indian Nations Who Once Inhabited Pennsylvania and the Neighboring States, 1819. American Philosophical Society, Philadelphia*

An unknown artist painted Salem from the northwest, c. 1790. The community was already characterized by the deep narrow back yards also common in towns in Pennsylvania German Country. Garden beds are clearly visible within the board-fenced back yards. *Private Collection. Photo Courtesy: Historic Salem, Inc. Winston-Salem, North Carolina*

Not until 1824, when the barn-like doors were converted into sliding glass doors by William Strickland, would the hall become a conservatory, but serious work on the garden was undertaken almost at once. A "fine garden" was wanted. Initially Reuben had a carpenter construct a new "Arbor in the Garden" and they hired a gardener named James to help with the grounds. Within a few years the arbor had "... a coral Honeysuckle on one side and a monthly [repeat blooming] rose on the other." The weed-choked garden had been reclaimed. In 1818 Jane wrote to Reuben from her family home on Long Island:

> ... I feel quite a regret that I cannot have a little oversight of the garden—but I do not doubt it will be pleasanter than it ever has been to me—James will not allow us at least to loose one another among the weeds—or suffer them to grow so luxuriantly as to become a hiding place for stolen goods—which thee recollects was the case there years ago—I hope to have honeysuckles and roses planted round the Arbour and Lilacks Snowballs and double Altheas at the end of the new path—if convenient a larch tree somewhere and the Bigonia radicans by those old trees in the front yeard—S. Johnson said last summer she would give me some roots of both common and variegated periwinkle—in the desk I left some poppy seeds that I gathered for sister SHB if these are wanted please take part and send the rest to me ...

Clearly the garden was not only being reclaimed, it was being changed from a kitchen garden to a mostly ornamental flower garden. Jane and Reuben, however, had different motivations for the change. Jane was an aesthetician. She loved color and beauty. Their wedding china was colorful botanic ware hand-painted with different varieties of flowers and vegetables. She ordered bright orange moreen bed hangings and she loved vivid roses, poppies, and pink honeysuckle. She wanted a beautiful landscape.

Reuben, in contrast, saw the garden as an extension of his major interests—natural history and agriculture. "He was as proud," it was observed:

> of his cabbages, fruit trees, and Glycene Apios (ground nut) as he was of the many rose bushes he and Jane planted. Besides his general interest in the cultivation of the earth, Reuben was an avid student of natural history and a close friend of such early 19th century naturalists as Thomas Nuttall, Thomas Say, Rubens Peale and his brothers Franklin and Titian, and John James Audubon. Sometimes alone and sometimes accompanied by one or more of these friends, Reuben traveled through the countryside collecting unusual plant specimens which he then pressed and incorporated within his herbarium, still housed at Wyck. Similarly, he swapped seeds and cuttings with friends in Germantown and with acquaintances from New England to the Deep South.

The Carolina allspice (*Calycanthus floridus*) was described by Moravian botanist Samuel Kramsch (1756-1824) as having a "fragrance of strawberries" after he moved from Nazareth, Pennsylvania, to Salem, North Carolina. When introduced north, the shrub's mahogany colored flowers became favorites of Pennsylvania Dutch girls, who tied the fresh blossoms in their handkerchiefs like sachets. These flowers, and one of the last year's seedpods, were photographed in a Lancaster County garden. *Photo: M. Susan Richman*

The leaves and flowers of the tulip tree (*Liriodendron tulipifera*) were sketched by botanist Louis David von Schweinitz (1780-1834), the American-born grandson of Moravian leader Count Nicholus von Zinzindorf. *The Academy of Natural Sciences of Philadelphia, Ewell Sale Stewart Library*

Reuben also appreciated the use of nature as an educational tool, especially for his own children. He wrote a very tender letter to Jane about this in 1818.

> ...with what pleasure will I not welcome thee to the scene of my present interesting occupations, put under thy tender charge the garden and see thy taste displayed in every shrub, and in the light festoons of every vine trace the hand that gave them grace. With what additional pleasure will I not aid thee in imparting instruction to our daughter and after the daily task has been cheerfully performed lead her forth into the garden and the field & teach her to admire the economy of vegetation in the bursting forth of the pumula and radioles of the legumes her little hands have deposited in the earth, point out the bees as examples to stimulate her industry, and see our mantle decked with the wild flowers which she has assisted to gather. But this picture impresses too deeply the privations thy absence occasions to thy attached friend and Husband

Their daughter and subsequent family members who lived at Wyck learned their lessons and loved the place. They updated it with lush Victorian plantings and made it more refined during the Colonial Revival Period, and today a conscious restoration is underway to return the garden to the Reuben-Jane period without destroying valuable specimens added later.

> In Reuben's eyes [wrote Sandra MacKenzie Lloyd], it was essential that parents teach their children to nurture the earth and make it fertile with vegetables and flowers. His philosophy corresponded with Quaker beliefs that the earth was a gift from God to man. Man, then, had the responsibility and pleasure of cultivating the earth for his own benefit and as a celebration of God's goodness. These tenets and their application to daily life helped make Quaker Philadelphia a center for agricultural and horticultural achievements during the eighteenth and nineteenth centuries.

The Pennsylvania German tradition was elaborated upon when Reuben Haines maintained extensive vegetable gardens and cropland at Wyck. The calm horticultural beauty of Jane Bowne was more Quaker inspired. Beauty and utility: Quaker and Dutch.

This study of fungi is one of thousands of sketches and finished drawings done by botanist Louis David von Schweinitz (1780-1834). This was an illustration for his pioneering *Synopsis Fungorum Carolinae*, the first important work published on fungi in America. Von Schweinitz is known as the "Father of American Mycology." *The Academy of Natural Sciences of Philadelphia, Ewell Sale Stewart Library*

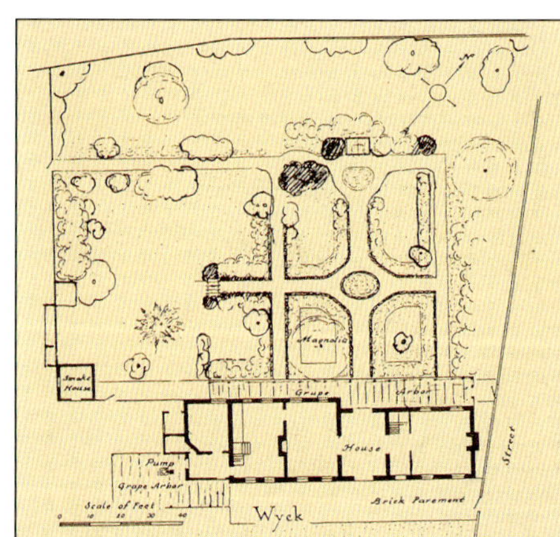

The plan of the house and the principal garden area of Wyck shows how they are beautifully integrated. Note how the conservatory opens to a grapevine-covered piazza, which in turn opens to the main axis of Jane Bowne Haines' flower garden. *American Philosophical Society, Philadelphia*

Flowers were highly prized by the Moravians. By the mid-nineteenth century the four square garden at the John Siewers House of 1844 had become a lively flower garden featuring a lush display of mid-summer annuals, including Joseph's Coat (*Amaranthus Tricolor*), flowering tobacco (*Nicotiana orlata*), hyacinth bean (*Dolichos lab-lab*), purple flowered petunia (*Petunia violacea*), cockscomb (*Celosia eristata*), and Lantana (*Lantana montividensis*). *Photograph by Virginia R. Weiler. Courtesy Old Salem, Winston-Salem, North Carolina*

The front of Wyck is set off by a tree-shaded lawn. Roses and clematis grow on the trellis on the house. Note the large potted plants brought out for the summer. In winter they are kept in the conservatory, through the sliding glass doors at the center of the house, as designed by architect William Strickland in 1824. The photograph is one of a series of stereopticon photographs taken by local photographer Harry Lewis in 1871 or 1872. *Wyck, Philadelphia*

Beans and squash and assorted herbs, including dill, share space with colorful flowers in a recreated garden near the First House of 1776 (itself a reconstruction) in the Moravian community of Salem, North Carolina. The flowers include Bee Larkspur (*Delphinium elatum*), heliotrope (*Heliotropium aborescens*), lavender (*Lavandula augustifolia*), Flanders poppy (*Papaver rhoeas*), pot marigold (*Calendula officinalis*), hollyhocks (*Alcea rosea*), and phlox (*Phlox paniculata*). Note the picket fence bounding the garden. *Photograph by Virginia R. Weiler. Courtesy Old Salem, Winston-Salem, North Carolina*

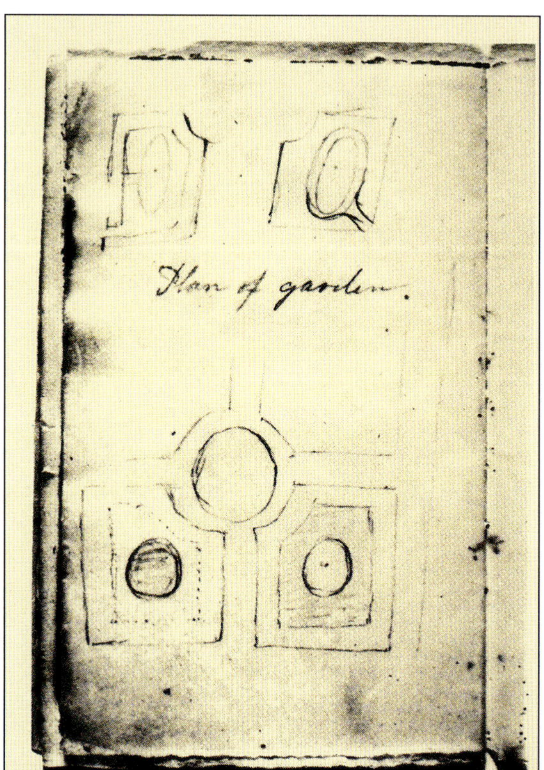

Jane Bowne Haines sketched her "Plan of garden" in her garden notebook, 1821-1827. It would become the basis of one of the oldest documented gardens in America, that at "Wyck". It is essentially a four square garden where flowers, and especially roses, replace vegetables. *Wyck, Philadelphia*

The garden side of Wyck is shaded by grape arbors. This 1871 or 1872 photograph shows the garden as lushly Victorian. *Wyck, Philadelphia*

Gardens are continually changing. Boxwood grows; other plants also mature and occasionally die. Two late-twentieth century views of the rose garden at Wyck demonstrate this very well. The roses, shown in June bloom, are almost all descendants of the original ones planted by Jane Bowne Haines. The paths are covered with pea gravel in nineteenth century fashion. *Wyck, Philadelphia*

At Wyck, an ancillary farm-related building provides a background for more informal, later gardens, but roses are still prominent. *Wyck, Philadelphia*

III. Landis Valley: Unique in America

Landis Valley Museum is sited on what has been an agricultural locus from the eighteenth century into the twenty-first century. Located three miles north of the City of Lancaster, in Manheim Township, Landis Valley is surrounded by some of the richest farmland in America, indeed, in the world. This is an area where agriculture continues to thrive and at the same time, paradoxically, continues to shrink because of suburban development. The story of the Landis family, their farm, and the museum they founded is by turn typical and unique. It is at the heart of the story of Pennsylvania German agriculture.

Open the 2004 Verizon telephone directory for Lancaster County and you find two pages of Landises. Mostly they have normative American given names – Charles, David, John, Mary, and Dolores, but there is also a Phares, an Elvin, and a Myron. Many Landises are in farming and in food-related businesses, and others offer traditional services like J. Omar Landis the auctioneer. Auctioneers have been a fixture of the Dutch countryside since the eighteenth century. Other Landises offer more contemporary services. Need dazzle, call Landis Neon Signs; need to rent a limousine, try Landis Luxury Coaches whose motto is the very 'un-Dutch', "You Deserve the Finest." Preserve your wealth with Charles D. Landis, Chartered Financial Consultant. If you are "chronologically challenged," decline gracefully at Landis Homes Retirement Community.

The Landises of Lancaster County are what this author likes to call a geographic family. Not only are many still locally centered, but their names are attached to many places ranging from Landisville to Landis Valley Road and the even more particular "Jake Landis Road." Read genealogies and you find intermarriages with other geographic families including the Dillers (Dillersville Road, *etc.*) and the Rohrers of Rohrer's Mill and Rohrerstown.

In origin, the Landis family is Swiss Mennonite. Records show them as living about twelve miles south of Zurich as early as 1438. The Steigerwalts discussed in Chapter I were Lutherans – Church people. The Landises were Sectarians, or Plain People. Some still are; others have become Church people. Some are unchurched. While the Church people, as noted earlier, came to America primarily for economic advantage, the sectarians came firstly for religious and physical survival as well as economic betterment.

Various Landis family members and their neighbors founded a Mennonite Congregation called Horerberg in Switzerland and Hans Landis became the first of his name to become a minister. Hans and his family would pay a dreadful price for their beliefs. Preaching throughout his Swiss Canton and the upper Rhine Valley, he was im-

A 1940s postcard shows an idealized vision of farming. Everything shown here was part of the Landis life. Mother Emma raised chickens. Father Henry H. and wife Emma built a new farmhouse, which they carefully maintained. Everyone went on picnics and the kids all fished. And brothers George and Henry K. went to innumerable auctions, gathering materials for their once and future museum. *The Cottage Collection*

prisoned and then, finally, executed at age seventy. He was the last Mennonite to be martyred by beheading in Zurich. Today he is commemorated in a hymn still sung in Mennonite churches. His wife would not renounce her faith and had her lands confiscated and was imprisoned – where she would die. Eventually, in about 1660, several of the Landis families moved to the German Palatinate where there seemed to be greater tolerance towards Anabaptists. However, this was illusionary; imprisonments, torture, and executions followed them. In 1717 three Landis brothers, Felix, Jacob, and John, along with their families and other Mennonites from their circle, made the decision to emigrate to Pennsylvania. John and his family would settle in Bucks County. Jacob and Felix and their families would establish themselves in Lancaster County and most of the local Landises are descended from them.

Lancaster County would attract many additional Mennonites and Amish in the early eighteenth century. The soil was rich and the terrain reminded them of home. Best of all, they were largely left alone on this frontier and were free to practice their unorthodox faith.

A few miles from what would become the center of Landis settlement, fellow Mennonite Christian Herr would build, in 1719, what is now the oldest surviving house in Lancaster County. Commonly known as the Hans Herr House, this large stone house originally served, not only as a farm center, but also a meetinghouse for the local Mennonite congregation. Two pages of Herrs are listed in the contemporary phonebook. Herr's Potato Chips, still owned by descendents, are a regional favorite. Dr. and Mrs. Donald M. Herr, both contemporary local veterinarians and antiques collectors have been strong supporters of the Landis Valley Museum.

In 1728, another very successful Mennonite farm family, the Johann (John) Snaveleys, would build another still surviving large stone house just down the road from what is now Landis Valley Museum. It is a building that was to play an important role in Landis history. The listing of Snaveleys in the modern telephone book, however, only fill a column and a half.

Now enter immigrant Jacob Landis who is affectionately known as "Pioneer Jake." We don't know when he was born, but he died in 1730 and he and his wife had three children who lived to adulthood. Their only son, Benjamin, married Anna Snaveley in 1749, the only child and heiress to the very prosperous 200+ acre Snaveley farm. By 1751 Benjamin and his wife owned the Snaveley family farmstead. Hardworking and lucky, the family prospered and by the time of Benjamin's death he owned 905 acres in the vicinity. In addition to farming, Benjamin was apparently a distiller and owned an interest in a mill as well as other agrarian related enterprises. Benjamin and Anna had a proper-sized family of nine children – four daughters and five sons. The middle son was named Henry and, because as a teenager he drove a Conestoga wagon to Philadelphia taking family goods to market, he became known in the family as "Drover Henry." Nicknames remain very important in the Pennsylvania German tradition where usually only a few family-related given names were used. This continues today among the Amish of Lancaster County who only have eight surnames and a very limited number of given names. For example, it would not be uncommon to find six or so men named Jacob Stoltuzfus living in one locale. One would be known as Jake, one as Shorty Jake, Red Jake might be a redhead, *etc.*

Drover Henry grew up to be the most successful of Benjamin's children and one of the richest men in his locale. He would marry three times, successively to Anna Long, Anna Stauffer, and, lastly, Esther Binkley (of nearby Binkley's Mill). All the wives came from geographic families. In total there were sixteen children. The youngest child was a son named Henry Harrison Landis who according to the custom of the time was expected to take care of his elderly parents. His reward was to be that he would inherit the family farm. It was a deal that Henry Harrison would regret. On March 3, 1876, he wrote in his diary, "...I done so an awful Business I got into by coming into this house if I had to do the thing again I would never come here if I was paid $3000 a year." Drover Henry had always been a demanding taskmaster and in the last year of his life he was demented and incontinent. "Just as I went for dinner," Henry Harrison recorded the next day,

> Pap shit his pants and shoes and the floor I undressed him and dressed him and washed him clean and when I was through with him he pissed on the floor. I cleaned that up then I could go for Dinner but such a smell will mak[e] you loose your appetite.

Henry had to cope with his father's problems and his mother's health largely alone. He apparently got very little support from his wife, Emma Diller [1842-1929]. Their marriage was, from his perspective, hellish and unhappy but it did produce four children. "...my wife," Henry wrote," ... is a perfect Devil and if she does not change or let up ... I am going to say good [by] to the whole shop." They remained unhappily married, however, 'til death did them part.

After his father's death, Henry was further frustrated to learn that he had not been bequeathed the farm, but only a life interest in it. It would go to his children on his death. This enraged him but he believed that he had to accept it – and he did. He farmed and ran a lime-burning operation and always he schemed for a big break that would provide him a lot of money – to show Emma that he was a success – but the big break never came.

> the first thing this morning was an Eruption between my wife and myself about money matters she is becoming mean in Spirit and decidedly disagreeable in her language. Just as usual when money is short she gets mean (Henry's Diary April 22, 1889)

By turn he speculated in grain and tobacco – often losing money. He became an agent for the new phosphate fertilizer and he would also buy lottery tickets – which were expensive for the day. "I am sending one Dollar to New York this morning for a chance in the little Havana" (Sunday, August 12, 1888)

> ... I am down in the dumps. I have nothing left but my hopebacked up by a determination to be somebody or something yet it seems rather in life to commence, but with wisdom and experience life and some I am able to pull along to some purpose I think I feel my insignificance at home here very keenly although none of the folks try to show it to me in fact there are none that come so plainly in view as the sarcastic view my wife taks of my opinions she sets her head regardless of my opinion or my wishes sit seems she tries to circumvent at every turn she seems to think I am done for she has my affairs and will do as she pleases without respect to any feelings I may have in the matter ["Whitsentide Sunday, June 9 1889]

Henry's immortality has come through his children – and their museum – and his diary record of his residence in and near the village of Landis Valley, which not only describes his personal life but is a detailed record of farming in the late nineteenth and early twentieth centuries, which was, in many ways, the zenith of the traditional German family farm.

Aside from the children, Henry's chief comfort was his mother, who died May 6, 1881. Friends, relatives, and neighbors gathered to pay their respects and to help:

> Weather rainy cool clear this eavening
> May Friday 6 1881
> Rose at one O clock sat up with mother until Breakfast Mother died at 7-10 this morning she expired without a pane or a struggle she is to day 83 years 5 months & 7 days B L Hefs came here in time to see her die B L Landis B L Hefs P R Landis came here and helped to plan for the funeral J K Hostetter Stephen Weller Ben Barnes John Lawrence were here & working for the funeral this eavening Annie & Levi Reist came here they behaved them selves first rate George Arner came here this eavening and staid until 9 Billy Whitcraft warned down at Dillers and brought Emma sister along up Mother look quite Natural this eavening a little full in the face
>
> weather fine this morning & after
> May Saturday 7 1881
> At home until noon house upside down women buisy [sic] everything in an uproar J K Hostetter is cooking meat down behind the smoke house Mrs Arner and Mrs. Fry are baking pies Annie Diller Kate Carper Mrs Steigleman ar picking out raisins William & John Lawrence and Billy Whitcraft are working about the Barn fixing up Tables &c&c this after I Emma & Annie went to Lancaster I bought myself a suit of Black clothes Emma got a crape Bonnet we came home about 3 Mother is lying stiff and cold in the Blue room I don't feel at home anymore since mother is dead
>
> fine weather all day
> May Sunday 8 1881
> at home the whole day I and Ladd went on the Hill this morning and fixed the fence in order to put the Cattle there tomorrow for the funeral there were people here on and on the whole day and especially this eavening there were quite a number here John Brubaker Michael Rohrer B L Landis I S Landis J K Hostetter Geo Arner John Carper and many others mother looks like life her corpse looks better now than when she was living, we keep a lamp burning in her room every night but we don't wake I feel awful lonesome when I lay down to sleep I miss Mother more than I can tell
>
> fine weather all day
> May Monday 9 1881
> The Baker from Town was here before we were out of Bed Billy went for Mrs Fry and John Carper went for Mrs Boly. Hollinger came with the Coffin about 6 o clock we took Breakfast about the same time soon after Breakfast the people began to gather in they sang a hymn at the house here and then started for the meeting House Hostetter Bomberger and Brubaker preached the latter in English after the burying we all returned to the house took dinner and soon after the people began to scatter and about 3 1/2 Emma pap and her Mother and Annie took leave they were about the last to leave I paid Mrs Fry $2.00 this evening

The one plus side of his mother's death is that Henry inherited some of her stocks and money, which helped shore up his often shaky finances.

Henry and Emma's family consisted of a daughter who died in childhood and three children who survived to maturity: Henry Kinzer (1865-1955), George Diller (1867-1954), and Nettie May (1879-1914). Nettie May was the darling of her family. She was doted on by her parents as well as by her much older bachelor brother, Henry Kinzer, who would be the driving force behind the museum. Henry also wrote a fragmentary autobiography that provides insights into this typical/atypical family. Among a people known for their piety, the Landises were very secular.

"One of the features of the writer's early life which was unusual," Henry Kinzer recorded:

> was the diversity of interests and opportunities which prevailed. His parents were not strict, but unusually liberal. They were not connected with any church, although the grandparents on one side were Mennonite [Landis] and on the other side they were connected with the Reformed Church [Diller]. He attended the Moravian Church while at school, and throughout his subsequent career his church attendance included many years of singing in the choirs of perhaps a dozen different congregations without partiality to any of them not other connection with any church. This same liberality was evidenced in the studies and reading, and consequently the natural bent of the mind was followed and led to the investigation of the unusual and the development of a judgment along the lines on independent thinking and wide reading.
>
> The parents of the author never belonged to any church, nor did their children. However, grandfather Landis and his wife were Mennonites, although not very active in the congregation, so that on several occasions he [Henry Kinzer] attended the old Mennonite church [still] at the corner of the farm, and the memory of those congregations is interesting, as their customs and dress differed somewhat from those observed in the present day. One saw the long waistcoat, bull-fall pants, boots, shad-belly coat with military collar, broad-brimmed hat with a straight high crown in which the brown silk handkerchiefs were kept. Such was the dress of the elderly people. The women were white caps and shoulder shawls, and were dressed in black. The women sat on one side of the room and the men on the other, and there was a stove between. In the front of the room was a large table around which sat the trustees of the congregation. Upon a low and narrow platform stood a bench against the wall, and upon this sat the preacher, and there was no pulpit. The singing was intoned in the usual manner, two lines at a time, since not all the congregation had singing books. These were the old

The exterior view of the Snaveley house of 1728 is remarkably little changed from when Henry K. Landis photographed it early in the twentieth century. The center of a prosperous farm, it came into possession of the Landis family via marriage by 1751. The date stone "1728 HIS" refers to the builders Johann and Hannah Snaveley (in the eighteenth century 'I' and 'J' were interchangeable). The house was "Rebuilt by Henry & Catherine L. Landis A.D. 1881. At that time the original stone section (left) was remodeled and its center chimney removed. A new frame wing (right) was added to the original house, as was the front porch. Note also that in the Henry K. photograph the house had fashionable modern foundation plantings as well as a small front yard flower garden. *Landis Valley Museum Archives; Landis Valley Museum; Craig Benner, Photographer*

style hymn books without notes and in prose paragraphs. The sermons and hymns and Bible were all in high German, for the Pennsylvania dutch was used only in conversation. One of the interesting things was to see the old preacher Carl Hostetter coming up the road in his high-backed sleigh with the off-side shafts, driving a fat horse, and with a heavy scarf around his neck ant the top of his high hat just appearing above the top of the sleigh. These high-backed sleighs were being displaced about that time by the more modern and ornate cutter. It was about this time that the customs of the Church were changing into a more modern trent, although the Amish retained their old customs and dress.

Henry Landis, Emma, and the children occasionally went to church but seldom, if ever, together. It was just part of life. Something to do. On a spring Sunday in 1881 Henry Harrison records:

> ... had a fine shower last night went to church this morning Rev John Landis preached in the English Language and C Hostetter in German J K Hostetter came ... I put my cattle on pasture to day for the first left the mare and colt out today a little rain to night Just enough to lay the dust

There were also other casual church goings when novelty was promised. On Monday, June 10, 1889, Henry H.:

> .... Walked up to the hotel this morning sen Dan Hostetter he told me there would be preaching by a Strange menonite preacher at the church so came home got myself and Geo and went to hear the Sermon both in English and German the men came from Chester Counter Jacob shovelharrowed corn in the 6 acre this fournoon this after he plowed the balance of the Orchard and Shovelharrowed my Tobacco George and Martin Landis went to Lancaster the afternoon for pleasure

Pleasure was also tied to religion. Occasionally the family would go to camp meetings. One Sunday in 1888 "George went Back the country to dunker diping near Voganville": to witness a Dunkard baptism. The Dunkards practiced, as they still do, total immersion baptism out-of-doors.

In addition to their casual attitudes toward religion, the Landis family was set apart from many of their peers because of their belief in education.

The relationship to education among the Dutch has long been complex. Because of the strong ethnocentric religious beliefs of many, most Dutch were very devoted to the concept of parochial schools, as the Amish and some Mennonites remain today. When Pennsylvania finally passed a Free Public School law in the first half of the nineteenth century, the greatest threat to its enforcement (as to its enactment) were many Germans who, rightly, believed that public education provided in English would impact unfavorably on the preservation of their culture. Assimilationists throughout Dutch country felt differently, and there were already many German-oriented schools. Run by the Moravians, the first group to believe in educating girls, Linden Hall Academy in Lititz, for example, was founded in the eighteenth century and remains a fine, albeit English language, independent school for girls today.

The Dutch attitude toward higher education was also conflicted, yet the need among the Lutherans, Evangelical Reformed Churches, and the Moravians for a learned clergy was an important early impetus towards founding colleges. Dutch country is dotted with small independent liberal arts colleges whose roots are grounded in traditional religious beliefs. The need for secular schoolteachers led to the founding of Millersville Normal School (now Millersville University).

Emma Diller, the daughter of a very prosperous farmer, had attended Linden Hall Academy. Henry H's education apparently was in local public school. Their children were to get as fine an education as possible. After starting in local schools, both sons were sent to Lititz Academy as boarding students. Both attended Lehigh University, but only Henry would graduate. Nettie May was educated locally and then at prep school in Brooklyn, N.Y., near where Brother Henry worked and George often visited. She began studies at Millersville, but had to withdraw because of her ill health. Nettie May suffered from tuberculosis and the family tried everything to help her, including a trip to a dry climate. Alas, the outcome was inevitable. Henry H's diary entry is poignant.

> 68 @ 7 a.m.Rain
> Wea. Tues. May 5, 1913. Ther.
> Rose as usual took Breakfast and felt cheerful not knowing what was in Store for me about 9 A.M. a Telephone message came Stating that Nettie Died yesterday at 12 45 at St George Utah I was So overcome that I cannot think or do anything I have never felt so overcome before I knew the tears were rolling down my cheeks and Sometimes I could not Speak and Sometimes I did know what I was doing this after I had to lie down for 2 hours since then I am Somewhat more like my self But I Still feel so Sorry for my little Girl Oh what I would not give to have my Nettie here alive and well

Nettie, born a farmgirl, was groomed by her family to be a Victorian or, more precisely, an Edwardian lady. Her death cast a pall over the family, but ultimately it had little influence on the future history of Landis Valley.

Central to the story, following the parents death, would be Henry K. with brother George as his sidekick. A family trip to the 1876 Centennial Exposition, commonly called the "Philadelphia World's Fair," was an eye-opener for the farm family. To eleven-year-old Henry it was a phenomenon ... one which would affect his whole life.

> One of [my] interesting reminiscences is a visit to the Exposition in Philadelphia in 1876. Of course it was a revelation to a farm boy, since an opportunity was afforded to see representatives of nations all over the world and the things that they produce for the commercial markets. At that time there was a coreless [Corliss] engine on exhibition in the machinery building — probably one of the first of that type to be made, and as a boy it appealed so much that while the family moved along he remained by the engine, and when he looked about he could nowhere see his parents, so he went on exploring himself, and got up on the elevated railroad and rode around the grounds and finally came back to the machinery building, where he again took a position beside this engine, and that is where his parents found him. They were the ones that were worried. This is not the first instance of proclivities shown towards wandering, and to taking care of himself as a matter of course. Some of the souvenirs gotten at that exposition are today in the Landis Valley Museum.

Henry K, whose family called him "Lad," would go on to major in engineering at Lehigh University and have a very diverse and cosmopolitan career. "Lehigh University was an engineering school," Henry wrote,

> and as this was the subject in which the writer was most interested, he anticipated a very pleasant and profitable period, especially since he chose the 5-year course in Mining and Metallurgy which provided a great number of subjects, more than the other technical courses. As it turned out, this was a good thing, since the activities when taking over educational work in New York City were quite varied and therefore needed a wide range of education.

After a few years as a mining engineer and an educator in the West, Henry settled in New York City, working as an editor of technical journals – especially those involved with the natural gas industry. For relaxation, he owned a houseboat, which he kept near Port Washington on Long Island, and he also had a sailboat. He even purchased land in Port Washington with the aim of settling there. Henry, however, usually vacationed back at the farm and often visited on weekends.

Gregarious, Henry had an active social life and he had many interests. He sang in church choirs and played in an amateur orchestra. Beginning in the 1880s he became an avid photographer and we owe him a great debt for the many photographs he took of his family and the farms and countryside of his life. Also always interested in the natural sciences, he and George even kept nude photographic records of their aging bodies for about twenty years, beginning in their twenties. They called these their "Anthropometric Studies."

While George did well at school, the life of the mind was not for him. He stayed at Landis Valley and helped run the farm, and at various times he managed a road crew and was the head of the City of Lancaster's sanitation department. He was also a very avid hunter. As a fine marksman, he hunted locally as well as in Clinton County, Pennsylvania, and he also made several trips into the West. George had many of his trophies mounted. In June of 1914, Henry H. records that the State Veterinarian, "... came up here [because] he wanted to see Georges

The original stone house on the Drover Henry Landis farm is today the center of a farmette and, alas, it has been swallowed up by suburban sprawl. *Landis Valley Museum; Craig Benner, Photographer*

The Pennsylvania Germans gave America the Christmas tree. The Victorians and the Edwardians gave us the image of the ideal family. As we see, the Landises were photogenic but they were far from ideal. *The Cottage Collection*

mounted Elk head." Neither brother married and George, who was the shyer of the two, was locally looked upon as an eccentric. His love of cats did not help his reputation — in later years he kept about twenty felines.

Both boys enjoyed the outdoors and both were collectors from a very young age, starting with found objects – Indian arrowheads, bird eggs, and butterflies. George started to purchase pistols and rifles when he was fifteen, but these early acquisitions were directed at his sporting interests. He would later expand his collection to include everything from eighteenth century Pennsylvania Long Rifles to fishing gear, including nets and spears of the types used in local creeks and rivers. One March day in 1876 Henry H. noted that a relative had come for dinner and then went to a neighbors, "... to lift some nets. I believe they had a fishing party they did not come back until evening."

The Landis family, early in the twentieth century. From left to right, they are Henry Harrison (father), George Diller, Nettie May, Emma Diller (mother), and Henry Kinzer. They pose on their side lawn. Notice that, in Victorian fashion, there are clumps of flowering plants scattered around the lawn. The sword-shaped foliage to the right is probably gladiolus. To the left are globe thistle (*Echinops*) and peonies. This photograph was probably taken in late June. *Landis Valley Museum Archives*

By 1910 George and Henry were beginning to collect very actively. They attended local sales, often buying cheap lots, mostly in Lancaster but also in surrounding counties. This sale going had started years earlier. Henry H. recorded on Saturday, November 30, 1889, that, "George went up to B.B. Landis Sale he bought a lot of old Trash some he brought home and some not." On Saturday March 14, 1914, "Georg [*sic*] went to the public Sale at Roseville Charlie Landis Sold his Mothers things he bought a lot" Had his son's taste improved? "What the brothers bought were usually ordinary items that reflected a past that was disappearing." These were ordinary items that ranged from plows to mousetraps, millstones to fat lamps. "No doubt their purchase reflected both their interest in everyday Pennsylvania Dutch history and their middle-class pocketbooks." Old time collectors would have called them "nickel men." They often bought assorted lots of goods and ends that nobody else wanted. This is how they acquired many of their butter prints, baskets, and cookie cutters. Many objects that later generations would prize were often looked upon, by most, as little better than trash.

As collectors the brothers were more connected, in philosophy, to their older contemporary Henry Mercer Chapman (1856-1930) of Bucks County, a pioneer collector of iron, pottery, and almost everything archaic, than to Henry Ford or Henry F. duPont, the founder of Winterthur Museum. DuPont was more the connoisseur rather than an amasser and Ford was especially taken by the provenance of the items he acquired.

The Landis brothers, on the other hand, collected whatever and wherever they could. One informant, who knew the brothers, related that when an old barn or shed burned in their neighborhood, Henry and George would ask the farmer for permission to rake through the ashes for hardware and old nails.

Henry K. Landis was also a bibliophile who collected an extensive library including hundreds of volumes related to the Pennsylvania Germans, most of which remain on the property. Lost, however, is the brothers' extensive collection of books, often rare, relating to natural history, especially ornithology, but also covering their whole range of naturalist interests including entomology and botany. When the brothers' collection became part of the Commonwealth's museum system, the early staff did not appreciate all of the dimensions of the brothers' activities or their inter-relationships. Accordingly, large parts of the brothers' library were sold. A similar fate befell many of the Indian artifacts the brothers had amassed.

This same malady infected many other museum sites as well. When Dauphin County acquired Fort Hunter Mansion, a home lived in by many generations, the decision was made in the 1960s to sell all Victorian objects from the house, changing the possibility of accurate interpretation forever.

Carl W. Drepperd, the first director of what was then called The Pennsylvania Farm Museum of Landis Valley, was apparently elated when Henry K. died. He could then "purify" the collection and focus it, as he understood it, on Pennsylvania German and farm related objects. Important dimensions of the brothers' interests were to be obscured. Numerous books were sold in the autumn of 1954 to Walter Schatzki, a rare book dealer from New York City. Included were many works that related to interests that could be traced to the boys' childhood collections. Among them were several books about bird eggs, for example, "British Museum. Catalogue of Birds Eggs, London, 1901, 5 vols" and a number of books on ornithology in several languages such as, "Audebert-Vieillot Oiseaux dorés, Paris 1802, 2 vols." And "Der

Alten Fechteranfangliches Kunst, Frankfurt, c1531 (bound with six other pamphlets)." Among the wide ranging English language classics the brothers owned were "Michaux Nuttal [sic], North American Sylvia, Philadelphia, 1859, 5 vols.;" "Trimble, Isaac P. Treatise on the insect enemies of fruit, New York 1865;" "Elliot, D. G. New and heretofore unfigured species of the birds of North America, New York, 1869, 2 vols.;" and "Wilson, S.B. and Evans. Aves Howaiiensis – the Birds of the Sandwich Islands, London, 1890-99." (The above bibliographic information is quoted from Schatzki-Drepperd correspondence.)

The brothers, who had collected arrowheads from local farms as children, continued their interests as adults. Their local materials are preserved in the collections of the State Museum of Pennsylvania in Harrisburg. They appear also to have avidly collected Indian artifacts when these items were still easily available. Many were plains specimens, probably acquired by Henry when he lived in Colorado and other areas of the west and by George on his hunting trips. Drepperd offered seven boxes of "...SURPLUS AND DUPLICATE INDIAN RELICS OFFERED FOR SALE BY SEALED BID." The contents of "Box #6" is typical.

Posing in the yard about 1910 when Henry K. was visiting from New York City. Brother George was probably the photographer. Typically, Henry H. and his wife Emma are not paying any attention to each other – or even touching. Nettie May holds a plant. Henry K. and his parents sit on a garden bench with a grapevine, and probably a fruit tree in the background. Also note the August Lily (*Hosta sieboldii*) near the father's foot. *Landis Valley Museum Archives*

Henry Harrison Landis, George and Henry's father, was a successful, if reluctant, farmer who was also a genial gentleman. He is posed near the family farmhouse. *Henry K. Landis, photographer, Landis Valley Museum Archives*

On a summer day, probably a Sunday, early in the twentieth century. Nettie May, Henry H., Henry K., and Emma (left to right) pose informally on their side porch. George, no doubt, was the photographer. The downstairs shutters are closed; this was done in summertime to keep the interior dark and cool. The outdoor pump is to the right. Note the bench on the porch holding houseplants; one of them is an amaryllis (*Amaryllis Hippeastrum*) in full bloom. *Landis Valley Museum Archives*

A collection of bead work and other odd items.
Sioux mask from Rosebud, S. Dakota. This mask is full size human form, measuring 11" from skull cap to chin.
Two beaded belts – maroon cloth and metal backing.
Winnebago gauntlet cuff, beadwork in magnificent pattern.
Two Sioux leggings – rawhide – completely beaded in magnificent pattern.
Beaded apron patch (modern)
Beaded small switch (modern)
Beaded head-band – Sioux (old)
Pair of deerskin moccasins, decorated with blue and orange beading and red felt.
Sioux moccasins, covered with magnificent beading.
Amulet beaded in green and yellow, with fine dangles of red feathers and black glass beads.
Apache dance rattle (damaged) seeds missing.
Indian tambourine and drum stick.
Gorget – 3 parallel rows of cut and polished bone.
Small mat decorated with colored grass in magnificent pattern.

During the Landis brothers' lives, their collection never was static. They often sold items that were duplicates or that no longer interested them. Occasionally, when they really needed the money, the brothers sold more important items that were at the core of their collection. The same informant who remembers the brothers raking through ruins to recover iron, recalls that "... after a storm blew down a shed on my father's farm in the 1930s we went to the brothers to buy missing hardware." Revivalist craftsman Clyde Stacks (1918-1987), would often embellish older pieces with his painted variations of Dutch designs. He had learned the craft from an antiques dealer friend while he was a student at Franklin and Marshall College in Lancaster. Ed Long often took his young protégée to visit the Landis brothers at their farm, where Stacks could see and study historic pieces with their paint still intact. He also remembered that the Landises occasionally would sell pieces from their collection to Long.

The route to public ownership and the permanent preservation of the Landis Valley Site was convoluted.

Henry K.'s trips home and his collecting persuaded him to give up his plans for a home at Port Washington and to retire back home where he joined George in caring for their parents, both octogenarians by 1924. The farm had changed over the years. In the 1870s Henry H. had created a new farm out of part of his life tenancy and moved the family into a new farmhouse in the heart of the small crossroads village of Landis Valley. Additionally, he built the necessary ancillary buildings. The remainder of the farm consisted of lands on the other side of the modern Route 272, surrounding the original, and still existing, farmhouse complex. It was on these lands that Henry had a small limestone quarry and his limekiln – and these fields were often rented out.

When Henry H. died in 1926, the brothers inherited the farm and it provided them with capital to help support their "Barn Museum," which they had opened in 1925. Farm income was never enough so that the brothers increasingly supported their operation by selling off land, apparently to their farmer neighbors. By the late 1940s, they had sold off most of their surplus acreage. Some money went to develop museum buildings. Firstly they adapted the stable to this purpose and then in 1939 they built the Yellow Barn, which incorporated materials acquired when the barn on the nearby the Jacob and Elizabeth Landis farmstead was demolished. Ironically, the Jacob Landis farm is now part of the museum and it now features a recreated barn.

Like Henry Chapman Mercer and many of their generation of collectors, the Landis brothers packed their buildings with artifacts as their interests turned from farming to preserving the implements of farming and the

Emma Diller Landis is the perfect, if lean, image of the Pennsylvania German *hausfrau*, as she stands, workbasket in hand, on the back porch of her farmhouse. *Henry K. Landis, photographer, Landis Valley Museum Archives*

domestic artifacts of the culture that revolved around farming. These masses of objects were unlabeled. The brothers believed that the best experience for visitors was to have the collections shown by a knowledgeable guide — and who better than Henry? The museum's reputation grew – especially among the Pennsylvania German community, but money problems remained a constant concern for the aging brothers. A white knight came to their rescue in the person of a wealthy German-born Reading industrialist. President of Berkshire Knitting Mills, Gustav Oberlaender (1867-1936), who founded and headed the Oberlaender Trust for Better Understanding Between Citizens of the United States and Germany. He was also trustee of the Carl Schurz Memorial Association and the subsequent foundation that was created to honor Schurz (1829-1906) who had become a poster boy for American-German relations. Born near Cologne, young Schurz had become disenchanted by despotism in his homeland and, after taking part in the failed Revolutions of 1848, he fled to America. Becoming a citizen, Carl quickly became active in Republican politics, campaigned for Lincoln, and served as a general in the Civil War, later becoming a United States Senator from Missouri as well as a journalist.

The trustees of the Schurz Association, learning about the brothers' collection, saw it as a vehicle to introduce Pennsylvania German culture to a larger and, perhaps, a national audience at a time when the German image in America needed all the polishing it could get. This image had been tarnished during World War I and now the rise of Hitler was disquieting to many. The brothers needed help but, like all true collectors, they were hesitant to give up control of their collections. Eventually a deal was negotiated which would benefit both the collection and the brothers. The Oberlaender Trust, which the recently deceased Schurz Trustee Oberlaender had controlled, agreed to pay for new exhibition buildings. In 1940 the Trust incorporated the Landis Valley Museum. For their part the brothers donated about three acres of land, the original exhibition buildings, and part of their collection. In return, the Landis brothers were to serve on the nine-member board of the museum and act as the museum's curators. For the first time, a professional assistant was available: Dr. Felix Reichmann (b. 1899), a young Viennese emigrant who came to America in 1939 and was employed by the Carl Schurz Foundation as a librarian.

When the newly renamed Landis Valley Museum opened in 1941, Trust money had wrought a miracle. There was a new complex of stone buildings – a tavern, the gun shop, and two large sheds for displaying agricultural implements. Behind the scenes, there was money for curatorial salaries and general operating expenses. Despite wartime travel restrictions, Henry worked tirelessly to promote the site, even providing directions on how to visit the Museum by bus. Admission was free, as the Schurz Foundation wanted to counter anti-German feelings during World War II. Visitors, however, were encouraged to join the Landis Valley Museum Associates (which still exists) for two dollars a year. In the year that World War II ended, 1945, 5500 visited the museum; by 1947 the visitation rose to more than 11,000.

Money again became a problem as the Oberlaender Trust, heavily invested in German securities, began to run out of money after spending about $90,000 on Landis Valley. Almost as important was the fact that no one at the Trust, or the brothers themselves, had the professional experience to develop the site further. No one had the stamina and determination to catalog the collections or to consistently interpret them. The Trust and the brothers turned to the state government for help and, after years of negotiation, the Commonwealth of Pennsylvania took title to the land, buildings, and collections of the museum as well as additional land, buildings, and additional collections owned by the Landis brothers. The brothers, George then 85 and Henry 87, retained life tenancy of their family home, kept some of their collections, and received lifetime appointments as curators. The Pennsylvania Historical and Museum Commission (PHMC) would administer the museum, which in co-operation with the Landis Valley Associates still operate the facility.

New meets old. For son Henry K., as a young boy, the most exciting outing the farm family took was to the Philadelphia World Fair in 1876. Henry later wrote that the sight of the Corliss Steam Engine decided his future career as an engineer. In this engraving Emperor Don Pedro II of Brazil waves to the crowd. U.S. President U. S. Grant stands to his right. *The Cottage Collection*

The PHMC enlarged the purpose of the museum by renaming it the Pennsylvania Farm Museum at Landis Valley, acquiring more land (including some the brothers had sold), moving in historic buildings, and creating new-old buildings to enhance the vision of crossroads village life. A new farm with buildings constructed of logs was created to demonstrate farm life in the eighteenth century. The Jacob Landis farmstead and the Landis Brothers farm would concentrate on nineteenth century farm life. Gardens were recreated, or newly created along historic principles. Historic farm animals were introduced and the enormous Landis Valley collection, amassed by the founders and added to by many donors, is finally being cataloged and housed in new, appropriate, storage and exhibition buildings. An Heirloom Seed Project was founded in 1985 to provide historically correct seeds to museums and the general population. Today, farming and gardening are again central to the Landis lands, where agriculture had never entirely ceased. Today you can see the Landis collection in buildings, enjoy the fruits of historical research in period gardens and fields reflective of the eighteenth and nineteenth centuries. Across the road (Route 272) is encroaching suburbia. Here, too, is a failure of vision...a failure to acquire this original Landis farmland and preserve it from development. Up the road you see twenty-first century agriculture in its many varieties. You see Amish dairy farms and the large-scale commercial farm-retail complex that is the Oregon Dairy. You see cut flower farms, truck gardens, and roadside stands. You see gentleman farms and farmettes (shorn of most land) beautifully maintained. And the 1729 Snaveley House is still down the road towards Lancaster, now incorporated into a small office complex. It is adjoined by churches and even a Jewish Community Center. The Landis Valley Museum is clearly the armature for the rest of the book. It is important to any study of Pennsylvania German agriculture and horticulture.

Young Henry K. Landis poses with a young boy, possibly one of his many cousins, on the back lawn of the family home early in the twentieth century. The garden fence is in the left distance. A freestanding grape arbor stands at right. *Landis Valley Museum Archives*

A favored childhood pastime for many farm children was fishing. Here Nettie May (standing middle right) baits a hook. Her pole is bamboo and no doubt store-bought. She and her friends were probably fishing along the nearby Conestoga Creek. Her mother, Emma, in defiance of fashion, kept her daughter's hair close-cropped. *Henry K. Landis, photographer, Landis Valley Museum Archives*

Nettie May Landis (right) and a more elaborately dressed friend sit on a well-maintained rail fence with bountiful Landis family cropland behind them. *Henry K. Landis, photographer, Landis Valley Museum Archives*

Young Nettie May visits with a friend who lived in a traditional plastered stone farmhouse. Many Pennsylvania German farmhouses grew grapes along the southern wall on trellises between the first and second floors. The grape leaves provided shade in summer. In winter the bare vines let maximum sunshine hit the stone walls. (Also see Chapter V.) *Henry K. Landis, photographer, Landis Valley Museum Archives*

Nettie May Landis is pushed on a swing by a cousin near a picket fence in need of repair. A barn is in the background. Note that the swing, in timeless fashion, is suspended from a tree branch. The trade card from the Bon Ton Millinery in Lancaster City shows a girl on a similar swing surrounded by peach and apple blossoms. A grownup Nettie May admires apple blossoms at Landis Valley. *Henry K. Landis, photographer, Landis Valley Museum Archives; The Cottage Collection*

Nettie May Landis was brought up to be a proper Edwardian lady rather than a future farmer's wife. She loved flowers and gardened when her often-frail health allowed. She favored roses, which appear on her hat and on the china tray she decorated. China painting was a highly admired art to be practiced by young ladies and Nettie May usually preferred motifs drawn from nature. She decorated the pitcher with currants. Her family was also proud of her oil paintings, which they expensively framed in gold moldings. Most were still life images like the pansies. She, in all probability, painted the hunting dog as a gift for her elder brother, George, an avid hunter. *Henry K. Landis, photographer, Landis Valley Museum Archives; Color illustrations by Nadine Steinmetz*

Women's work within the Landis family included gardening and caring for the chickens. Chickens often had free run of the farmyard and they would nest wherever they wanted. And obviously, as seen in this regional view, not all farms were as well kept as that of the Landis family. One chore was to find eggs, as dramatized in the early postcard. A modern interpreter stands in a recreated garden at the Landis Valley Museum. *Collection of Mr. and Mrs. Michael B. Emery; The Cottage Collection; Landis Valley Museum*

The Landis family's 1870s farmhouse was photographed many times by Henry K. The view of the house from the road shows the surviving Lombardy poplars (*Poplar Nigra c.v. "Italica"*), which once adorned the front of the property. Note the grape arbor attached to the carriage house in the background. Early twentieth century views of the back of the house (before and after the extension of the summer kitchen) show rather informal landscaping with shrubs and perennials. *Henry K. Landis, photographer, Landis Valley Museum Archives*

The Landis farmhouse was built right at the crossroads of the village of Landis Valley. The farmstead was fronted with a carefully maintained picket fence and lined with short-lived Lombardy poplars. In the center is Hilton's auction barn. Henry H. often bought livestock there and made money pasturing horses brought in for sale. To the far left is the Landis Valley Hotel where Henry H. often passed the time of day and drank. *Henry K. Landis, photographer, Landis Valley Museum Archives*

The Landis family home as a museum house in autumn. The millstones were arranged on the side lawn by Henry and George after their parents' deaths. The conical brown sandstone stone was part of a hemp mill. *Landis Valley Museum*

The staid Landis family parlor was meant for formal entertaining and it clearly demonstrates that the family was fashionable. Even here a love of nature is reflected in the floral carpets on the floor. *Landis Valley Museum*

A well under an outside porch gave the Landis Family a convenient water source for family needs. The pump could supply water directly to the dry sink. *Landis Valley Museum*

The Landis brothers built the Yellow Barn as their first new Barn Museum structure using materials rescued from the barn on the Jacob Landis farm. The stone wagon shed was added in the 1940s. In this photograph by Henry K., a young knickers-clad worker rakes the newly planted lawn. *Landis Valley Museum Archives*

Every surface was used at the Landis brothers' Barn Museum, as seen in this 1930s view by Henry K. Some of the museum's collection of hardware and a few of its fine weathervanes were simply nailed to the walls of sheds. Daylilies (*Hemerocallis fulva*) grow in a stone cattle-watering trough. *Landis Valley Museum Archives*

Henry K. welcomes guests to the newly completed tavern building in the early 1940s. This rare hand-tinted photograph shows a potted geranium standing on a pedestal in the right foreground. In the modern photograph the tavern is adjacent to the sheathed log Erisman house, which was moved to the site in the 1960s. Landscaping has softened the vistas. *Landis Valley Museum Archives; Landis Valley Museum*

Henry K. Landis was very proud of his collections. In one of the 1940s photographs he displays a painted plank bottom chair, brightly decorated with floral motifs, which he holds next to a flower bed. The foliage of the hollyhock (*Alcea rosea*) is visible to the left. In the other photo he holds an unidentified object amidst his collection of millstones. The domed top stone resting to the left is a glaze quern – designed to grind lead oxide used in glazing redware pottery. *Landis Valley Museum Archives*

Henry K. also collected parasols. He favored those that reminded him of blossoms. These are samples of a large series he photographed in the 1940s. *Landis Valley Museum Archives*

George and a visitor pose in the 1930s on a bench in front of cast iron hitching and fence posts that were propped against a side of the house. Collections were displayed wherever there was bare space. Houseplants are visible in the window. *Henry K. Landis, photographer, Landis Valley Museum Archives*

George Landis stands by the pump installed behind the new Tavern building constructed in the 1940s. *Henry K. Landis, photographer, Landis Valley Museum Archives*

The Landis brothers collected many artifacts that were so large that they couldn't properly care for them. One of these was a cider press that Henry photographed in the 1940s, shortly after the stone wagon shed was built by the Oberlander Trust. Because of weathering, only the huge wooden screw, now safely housed, remains. A helper uses a reel-type push lawn mower. Note the vigorous castor bean plant (*Ricins communis*) to the left. It is an exotic that the Dutch have grown since at least the mid-nineteenth century. *Landis Valley Museum Archives*

The Jacob Landis farm, as it looked in the 1930s before its restoration. To the right, partially obscured by the trees, is the barn being demolished. Its timbers were used to construct the Yellow Barn exhibition building. *Henry K. Landis, photographer, Landis Valley Museum Archives*

The Jacob Landis farmhouse and the *Grössmütterhaus* after restoration are proudly set off by reconstructed picket fencing. The complex also has a fine outdoor bakeoven. *Landis Valley Museum*

Built by Jacob Landis, Jr., the Landis Valley Hotel was an active hospitality venue for well over 100 years. It was operating as "Louie's Bar" when the Commonwealth of Pennsylvania acquired and restored it to an earlier appearance. Needless to say, it no longer serves alcohol. Every crossroads town in Dutch country had to have three buildings – a tavern or hotel, a general store, and a blacksmith shop. Landis Valley has all three, but the hotel is the only one original to the site. *Landis Valley Museum*

The Swiss-style bank barn on the Jacob Landis farm was constructed in the 1970s to replace the one demolished by an earlier farm owner. The new structure was designed by architect John Heyl of Allentown to be representative of German barns. It is not based on any single prototype. *Landis Valley Museum*

The Log Farm at the Landis Valley Museum was built in the 1970s to portray mid-eighteenth to early nineteenth century farm life. Its major structures are a center chimney house and a *grundsheir* or ground barn. Designed as a living history complex, it features historically appropriate animals like the pigs resting in the pigsty and the horse being fed by a costumed interpreter. *Landis Valley Museum*

The "Homestead of Jacob H. Landis" was painted by Jacob Stauffer (1808-1880) in 1879 to commemorate the visit of the artist and a local physician to the very prosperous farm and mill complex owned by a distant relative of the Landis Valley family. The artist, his host, and Dr. Rathvon are seen on the front porch. The mill is to the left; the original farmhouse is next (note that it is on axis to the red barn). The newer, larger stone house is placed closer to the road to better display Landis' prosperity. Enclosed flower and vegetable gardens are shown on the right in front of the barn. These amazing buildings still exist along Route 999 near Millersville in Lancaster County. It is the most architecturally significant Landis related complex in Central Pennsylvania. Landis forbear Hans would be proud of what his descendants built. Unlike Landis Valley Museum, it is in private hands. *Watercolor and ink on paper, 9 5/8 x 13 3/8", Collection American Folk Art Museum, New York, Promised gift of Ralph Esmerian, P1.2001.63, Photo © John Bigelow Taylor, New York*

IV. The Pennsylvania German Farm

"Among the Pennsylvania Germans," writes author Dennis Boyer,

> ...there is a nearly three century-old love affair with farming that is only now starting to dim due to land prices and urban sprawl. Undoubtedly this love affair goes back to Europe — to the cantons of Switzerland, to the fertile plains of the Pfalz and to the green meadows of Alsace — stemming from a time when family land ownership was beyond the grasp of ordinary citizens.
>
> But for the better part of two centuries, to be "Dutch" was to be a farmer. Even for town-dwellers in the region the real family home label was assigned to an uncle's or a grandfather's farm.

Best-selling author James Michener did extensive research on the special agricultural area that encompasses Lancaster County. He adopts a global perspective in his massive novel *Centennial*, employing a nineteenth century persona.

> If in the year 1844 a group of experts had been commissioned to identify the three finest agricultural areas of the world, [he wrote] their first choice would probably have been that group of farms in the south of England, where the soil was hospitable, the climate secure, and the general state of husbandry congenial. Here sturdy farmers, well versed in the ancient traditions of the countryside, raised Jersey cattle, plump black-and-white Hampshire hogs, rugged Clydesdales, and poultry of the best breed.
>
> The experts might also have selected that rare and rich band of black chernozem that stretched across southern Russia, especially in the Ukraine. Two feet deep, easy on the plow, and so fertile that it needed less than normal manuring, this extraordinary soil was unequaled and constituted an agricultural treasure which the serfs of the area had mined for the past thousand years without depleting it.
>
> But whether the experts had chosen England or Russia, one area that would have had to be included was that fortunate farmland lying around the small city of Lancaster in southeastern Pennsylvania, at the eastern foothills of the Appalachians. For sheer elegance of land and profitability of farming, it stood supreme.

Losing prime farmland is an old story. Today there is just so much less left. This "Truck" or produce farm, called Lochiel Hill, was developed into 12th Street in Hanover, York County, beginning in 1893. This photograph was probably taken as a souvenir view in November 1892. John W. Matzinger leased the farm from 1883 to 1892. The description of the personae, recorded on the back, is unusually detailed. Reading left to right: "Sarah A. Mrs. Matzinger [inexplicably holding a broom] ... Hireling Frank Bowers, Prop. John Matzinger, in wagon Master Jos. Matzinger, Driving Hireling Charles Fry two Black Horses Big and Little Bull: Two Dogs Big Rover Terrier Sport." All were recorded in a "Picture taken by Hattie Kirk School Teacher Lochiel School" whose shadow appears in the foreground. Nineteenth century photographs by identified women are rare. *The Cottage Collection*

It was not flat land. There was just enough tilt to the meadows to prevent rain from gathering in the bottoms and turning the land sour. Nor was the topsoil unusually deep or easy to cultivate. If a man wanted to have a fine Lancaster farm, he had to work. But the rainfall was what it should be—about forty inches a year—and there was a change of seasons, with frosty autumns during which hickory nuts fell and snowy winters when the land slept.

Perhaps this region, the heart of the Pennsylvania German belt, was, indeed, the true agricultural Eden—and it was populated by the perfect people who would work this remarkable land.

The Lancaster farmer did not exaggerate when he boasted, "On this land a good man can grow anything except nutmeg." And on all of it he could make a handsome profit, for his farm lay within marketing distance of Philadelphia and Baltimore. Corn, wheat, sorghum, hay, truck, tobacco and even flowers could be got to market, but it was the animals who prospered most and provided the best income, particularly cattle and hogs. Lancaster beef and pork were the standards of excellence against which others from less fortunate regions were judged.

Mahantango Valley Farm, in upper Dauphin County, was painted by an anonymous artist, c.1860. It celebrates the Dutch love of order. The fields are precise, the fences are fine, and even the trees in the woodlot are in straight rows. *The National Gallery of Art, gift of Edgar, William and Bernice Chrysler Garbisch*

Not all German farms were picture-perfect and neat. Cloths dry on the front lawn of a Gothic revival style farmhouse with two front doors, built on axis with the bank barn. That structure has been enlarged to contain an open wagon shed, with storage above. A springhouse is in the left foreground. Other farm buildings are in the middle distance. Two boys show off their "motor power" in late nineteenth century Berks County. *Mr. & Mrs. Michael B. Emery Collection*

In the divine lottery which matches men and soil—that chancy gamble which often places thrifty men on fields of granite and wastes good farms on incompetents content to reap whatever the wind sows—a proper match-up was made. Into Lancaster County, in the early years of the eighteenth century, came a body of well-trained peasant farmers fleeing oppression and starvation in Germany. Arriving mostly from the southern parts of that country, they brought with them a rigorous Lutheranism, which in its extremes manifested itself as the Amish or the Mennonite faith.

Many farmers would leave an attractive shade tree growing in the farmyard – often a remnant of earlier woodland. Eastern Sycamore or Buttonwood (*Plantanus occidentalis*) like this specimen or White Oak (*Quercus alba*) were favorites. Note the large haystack in this Lancaster County farmyard. *Henry K. Landis, photographer, Landis Valley Museum Archives*

All of this is summarized the most lyrically in *Rewards and Fairies* by British author Rudyard Kipling (1865-1936):

> It's a kindly, softly country there, back of Philadelphia among the German towns, Lancaster way. Little houses and bursting big barns, fat cattle, fat women, and all as peace-ful as Heaven might be if they farmed there.

Culture has shaped the landscape of the Dutch country. While the region certainly has its share of the same strip malls and generic housing developments that deface and attempt to homogenize much of America, there are still substantial areas where the region's distinctiveness shines through, but you must be attuned to read the unwritten messages.

First and foremost, as suggested, the Dutch have been an agricultural people, and agriculture, as we know, remains the dominant industry in much of the region. The farms, their layouts, and their buildings are all distinctive. And in various counties, as we have seen, "Century Farms" (farms that have remained in single family ownership for 100 years or more) are noted and celebrated. There even are numerous farms that have remained in family ownership for over 200 years. In John O'Hara's best-selling novel, *A Rage to Live* (1949), the hero, Sidney Tate, comes to central Pennsylvania to buy a fine farm; he is soon discouraged when he learns that the best, "...usually stayed in the family." Sidney gets his farm by marrying into it through the person of Grace Caldwell, a scion of an old, if non-Dutch, family.

The Dutch are a conservative people who value tradition and stability. As you drive down roads in agricultural Dutch country, note that for the most part, houses and barns are built on an east-west axis with the house and barn facing south. Additionally, the farm complex usually faces a road or lane. If a thoroughfare is missing, be assured it once was there. It is theorized that the buildings are on axis because in Europe, houses and barns were often attached and, with more land available they simply separated. Indeed, there is evidence to suggest that the connected house-barn, *hofhaus*, was used in Pennsylvania. In what is believed to be the first house in Bethlehem, Moravian leader Nikolaus Ludwig von Zinzindorf (1700-1760) wrote on Christmas Eve in 1740 that he went into the next room "to tell the beasts of the birth of Our Lord," informing us that he was in a barn-house. Census records reveal that house-barns continued to be occupied until the late nineteenth century in some Pennsylvania German areas.

Fantastic postcards proclaiming a region's agricultural plenty were popular in the early twentieth century. Reedsville, in Mifflin County, has had a very large normative German population and a growing Amish presence. What is a better symbol than an extravagant cabbage in this sphere of "The Sauerkraut Yankees"? Sent to Helen Gonder of Dauphin. The writer observed, "One of these would make some cole slaw." *The Cottage Collection*

The farmscape at the Oregon Dairy just north of Landis Valley is dominated by a huge modern cow barn and enormous silos, all of which dwarf the old bank barn and the traditional farmhouse. As with many successful modern farming operations, Oregon Dairy bottles its own milk, makes ice cream, and operates a successful restaurant and supermarket on site. *Oregon Dairies*

Having buildings facing south is commonsensible – buildings facing south are easier to heat. Curiously too, many of the houses have two front doors. And while many farmhouses, as well as those in small towns, have been covered in some unsympathetic materials including aluminum or plastic siding, most basic house forms remain very identifiable.

Neatness and order is characteristic of Pennsylvania German farms as is the dominance of the agricultural, rather than the domestic, buildings. This, plus a sense of tradition, is noted again in Michener's *Centennial*, as the author describes the Zendt farm. "From a long distance," he writes, "the stately stone barn … was visible in the cold moonlight." The farm had been first settled in 1713 by Mennonites and "As with any self-respecting Lancaster farm, the barn was six times the size of the house, for Amish and Mennonite farmers understood priorities." Trees were prized for utility and as a statement of order. The farm is witnessed through the eyes of a young Zendt son in 1844.

The Zendt farm contained many fine trees: there was the perpetual woods first harvested in 1701 when Melchior Zendt arrived from Germany; then there was the line of trees along the lane, planted by his son Jacob in 1714, and best of all, there was the miniature forest set out by Lucas Zendt in 1767. It rimmed the far end of the pond and was as fine a collection of maple, ash, elm, oak, and hickory as Lancaster Country provided. Each tree on the Zendt farm was a masterpiece, properly placed and flourishing.

When he reached the farm buildings, the young man looked briefly at the huge barn, then at the small red building in which he worked [a butcher house], then at the even smaller one stained black from much smoke, then at the various snow-covered pigsties, chicken coops and corncribs. Finally, tucked in among the larger buildings, there was the house, a small affair.

Hay no longer is stored in barns. Wrapped in plastic, it punctuates the landscape from summer to summer. The long white roll lies in front of a tobacco barn. *Landis Valley Museum, Craig Benner, Photographer*

This same neatness and order was evident as the twentieth century begins as experienced in the writings of Helen Reimensnyder Martin (1868-1938), a college-educated Pennsylvania German woman whose *Tillie the Mennonite Maid*, first published in 1904,was to be a best-seller, going through twenty editions. As an enlightened young woman, Mrs. Martin was not entirely sympathetic to the Plain people who she saw as being hostile to modern women. Her heroine, "Tillie, is an intelligent young girl, who to the immense disgust of her harsh, tyrannical father, prefers to read novels rather than toil long hours in the celery field as obedient Mennonite children were expected to do." Mrs. Martin's view of the Getz farm where Tillie lived with her family is described precisely, but coldly.

> The Getz farm was typical of the better sort to be found in that county. A neat walk, bordered by clam shells, led from a wooden gate to the porch of a rather large, and severely plain frame house, facing the road. Every shutter on the front and sides of the building was tightly closed, and there was no sign of life about the place. A stranger, ignorant of the Pennsylvania Dutch custom of living in the kitchen and shutting off the "best rooms,"—to be used in their mustiness and stiff unhomelikeness on Sunday only,—would have thought the house temporarily empty. It was forbiddingly and uncompromisingly spick-and-span.
>
> A grass-plot, ornamented with a circular flower-bed, extended a short distance on either side of the house. But not too much land was put to such unproductive use; and the small lawn was closely bordered by a corn-field on the one side and on the other by an apple orchard. Beyond stretched the tobacco- and wheat-fields, and behind the house were the vegetable garden and the barn-yard.

The total number of farms in Pennsylvania in 2004 is 58,200. The majority of them are in the "Dutch Belt." Many, if not most, of the 5293 farms in Lancaster County display the traditional layout, often with modern additions. Reinforced concrete or metal silos are twentieth century replacements to the wooden ones that were added to the farm-scape after the Civil War. The buildings that proclaim a successful twenty-first century farm are not usually picturesque additions to the landscape. Metal sheds house the large scale, heavy machinery that modern farming requires. Old barns, on most farms, are only used for storage, except where herds are small or if the farmer keeps horses or mules. On many farms today, even hay doesn't enter the barn for storage, but rather is left in fields in huge plastic wrapped rolls. Hay bales, that symbol of rural America, since the late nineteenth century, are becoming obsolete. Straw however is still baled. Farms that raise poultry have huge metal chicken houses that are highly mechanized and shelter tens of thousands of birds. Barns for dairy herds are commonly metal sheds as well.

At Kreider Farms near Manheim the family brags about their, "awesome cow palace" where they " ... treat our cows like queens!" The dairy barn is so huge that modern tour buses can drive through it. The cows are mechanically milked on an "amazing milking merry-go-round – fifty-four cows ride it at one time for true efficiency." Now the manure pile is no longer a sign of wealth. The forty-two tons of manure a day that the farm produces needs to be disposed of – a whole new technology has evolved. As with many big dairy farms, the Kreiders bottle their own milk and even operate a small regional restaurant chain, which features its own ice cream. A similar farming operation at the Oregon Dairy, just north of Landis Valley Museum, has a bottling plant, an ice cream facility, a restaurant and a supermarket on site.

At the same time that agriculture is getting bigger and bigger, many Lancaster County farms are being turned into boutique enterprises where baby vegetables and greens are raised for upscale Philadelphia restaurants, or game or free-range poultry is raised for the same markets and other specialty outlets. The variety of the region's agricultural output is increasing constantly. *Philadelphia Magazine* recognizes that:

The Kreider Farm near Manheim boasts an "awesome cow palace" which is so huge that modern tour buses can drive through it. Fifty-four cows at a time are milked on "an amazing milking merry-go-round." Kreider Farms combines modern agribusiness, retail outlets, and what is being called "agritourism" (farming tourists). *Kreider Farms*

… all eggs are not equal: and at Meadow Run Farm in Lancaster County, … free-range chickens fed a mineral-rich diet produce the ultimate omelet ingredient: Substantial shells open with a satisfying crack to reveal a thick, rich orange yolk. On summer Saturdays, farmer Phillip Landis peddles them at the Rittenhouse Square Farmers Market …

Unlike many other agricultural areas, especially those in the Midwest, 99 percent of Lancaster's farms are family-owned and relatively small in size. Because of the rich soil, the seventy-eight-acre average sized farm could allow a family to prosper. By contrast, on a whole the average farm size in Pennsylvania is 132 acres. There are 411,848 acres of farmland in Lancaster County (7.7 million in Pennsylvania) and, therefore, 69 percent of the County is in farmland (as opposed to 27 percent statewide). Lancaster farmland is valued at $7955 per acre as opposed to $3100 statewide. Cash receipts for crops and livestock was $914 million, almost 25 percent of the state's total ($4.04 billion). The Farmland Trust boasts that each year Lancaster County feeds 11.9 million people their eggs, 10.2 million their milk, and 4.6 million their pork.

Milk is bottled at Perrydell Farm in York County, which feeds its Holsteins exclusively on farm-grown feed. On the forefront of successful boutique agriculture, the farm's hormone-free milk is available in "creamy chocolate," vanilla, and strawberry flavors as well as plain. Most is sold the same day it is produced at the owner family's farm market store. *Lancaster-York Heritage Region*

Modern egg production is highly mechanized. Eggs emerge from washing, ready for inspection and packing. The process is less colorful than the traditional techniques but much more efficient. It also allows the cost of eggs to remain low. *Kreider Farms*

Early boutique farming: The Allentown area was celebrated in the Dutch Country for raising ducks for the table, much as was New York's Long Island. Roast duck was a Pennsylvania Dutch delicacy. Visiting the duck farms was a popular Sunday outing for turn-of-the-twentieth-century excursionists. *The Cottage Collection*

In addition to direct value of farm products, "Every $1 of product leaving Lancaster's farms boosts our economy by $5.88, a value [of] more than $4 billion per year." In Lancaster County 51,289 people out of a civilian workforce of 255,400 is in agribusiness (1 in 5). Farms also enhance tourism, which annually attracts seven million visitors, who spend $1.6 billion a year. Tourists might be attracted by the farms and the Amish, but the services they require in the modern world have led to much strip development in the heart of some of our best farmland. However, Lancaster County is number two in America in farmland preservation with 18 percent of its agricultural-zoned land, 58,000 acres, preserved.

No matter the changes the modern era has brought, many of the markers of the historic era remain – in farmscapes and buildings.

Like the Swedes and the English who preceded them into Pennsylvania, the Germans brought with them notions of construction and space usage from Europe. Early houses were half timbered, log, or masonry. The most common floor plan of all of these houses was a three-room plan common to much of central Europe, but used almost exclusively by Germanic settlers and their descendants in America. The *flurküchehaus* was customarily divided into three living spaces. The largest of these was the kitchen, or *küche*, which occupied approximately one-third to slightly less than one-half of the total floor plan. The front door of the house, very often a double articulated, or "Dutch door," generally opened into the *küche*, which traditionally ran the entire length of the house. In very rare instances, the rear of the *küche* is divided off into a small chamber or *kammerle*, which could be used for sleeping or storage, or both. The staircase to the loft above or the second floor was usually narrow and encased and was placed along the wall to the right of the doorway and opposite the fireplace. The interior cellar stairs, if any, were usually located under these stairs. There might also be a bulkhead door leading to the cellar from outside. A large stone fireplace was placed in the center of the room's left hand wall. Customarily, the entire stack was in the kitchen. On the chimney wall and usually facing the staircase was a door leading into the stove room, or *stube*, which usually occupied about two-thirds of the house's remaining space. Opening from the back of the *stube* was the bedchamber, or *kammer*.

The mushroom industry in America was founded by Quakers in Chester County. The most important growers and distributors are still the Quaker Phillips family, but there were also many smaller German growers as well. Grown in dark "Mushroom Barns" in a mixture rich in horse manure, locals often call white mushrooms "shit blisters." Like much in agriculture, the region is changing. Many growers today are Italian or Mexican in origin. In the 1950s, when this card was issued, the white mushroom was king. The region now produces many exotics as well. *The Cottage Collection*

Growing high quality wine grapes is a new form of agriculture in Dutch Country, which now boasts numerous vineyard wineries. These Seyval Blanc grapes were grown at one of the region's first wineries, Nissley Vineyards in Bainbridge, which was established in the 1970s. *Nissley Vineyards*

Washington Boro in Lancaster County prides itself on the earliest and best tomatoes in the region. Amateurs and professionals compete to have the first ripened fruit. The plants, started in greenhouses in January, are transplanted to the field by April, protected as needed by plastic tents, where they are staked and suckered. Town residents claim that the rich sandy soil, occasionally flooded by the nearby Susquehanna River, and the warm, humid climate give the tomatoes their perceived special flavor. *Lancaster-York Heritage Region*

The *küche*, or kitchen, was the food preparation center of the house and more. Traditionally, it was the only room in the house where a fire was kept burning continually. In winter, it was often the only warm room in the house. The large hearth openings in many German kitchens are often misunderstood today. Huge fires were never built in these fireplaces; rather, part of the hearth was used like a modern range. Several fires would be kept going for the requirements of what was being cooked: a lively fire for boiling a large pot, a few glowing embers to keep a small pot's contents warm. Part of the large fireplace was used to store several days' worth of firewood. The dry heat from the hearth fire would complete the drying process so that the wood was ready to burn. In some large fireplaces, caged hens were housed to provide the family with a few fresh eggs. Odorous, of course, but people had sensibilities different from those of our deodorant enhanced times. A very important ancillary part of the kitchen was the bake oven, which was usually constructed outside the house, convenient to the kitchen entrance. A small open shed kept firewood dry and protected the outdoor baker.

Water was very important to a household. The most desirable water source was a nearby reliable spring or a dug well. Often buildings would be constructed near a channeled spring. These "springhouses" were used to store perishables as well as to protect the water source. Butter making was a significant revenue source for many farmers or farmwives, especially those who lived near sizeable towns like Lancaster. Henry H. Landis often was his farm's butter maker, although it was more common as women's work. A springhouse provided just the place to allow cream to rise from the milk. A springhouse was important for making some cheese as well. German Swiss houses were often built over a spring – the spring ran through the basement. This provided an indoor water source and a primitive refrigerator. A special insulated ceiling made of mud and straw wrapped laths kept the cellar cool and helped to keep the upper rooms dry. Owners of German Swiss houses today usually have channeled the spring away from their houses or run the spring through culverts. If there was no spring, a farm's water source would be a hand-dug well, which would be topped with some device for drawing the water. Some Dutch farms used buckets lowered on cranks to draw water or, more commonly, they used hand pumps. Today, when few people use functioning hand pumps, except some of the Plain people, farms often retain their pumps (or have restored them as "old-fashioned" ornaments).

Sunflowers grown for both seeds and cut flowers are a very attractive crop with a long lineage in Central Pennsylvania. *Landis Valley Museum; Craig Benner, Photographer*

Amelia (1811-1877) and Eliza (c.1809-1886) Danner were painted by Jacob Maentel (1778-?) near Hanover in York County c. 1815. Fashionably dressed, they are nevertheless engaged in a childhood chore, feeding chickens. In the distance are both their Germanic end-chimney log farmhouse and a fine stone barn. *Watercolor, gouache, ink and pencil on paper 10 1/2 x 8 1/8" (sight), American Folk Art Museum, New York, Promised gift of Ralph Esmerian P1.2001.9*

Always important was the well maintained cellar. Cellars were usually dug under only part of the house. A dry, cool basement was important for storing many foods over the winter. If there were a spring, food storage areas would be carefully walled off, away from the water. Most cellar floors were hard packed earth, although some were paved with brick or stone. Characteristically, many German cellars contain large arches that support the massive chimney above. Many food storage areas had plaster applied directly to stone walls and were then regularly whitewashed to reduce vermin infestation and provide more light.

The physiognomy of the *küche* has changed over time. Many rooms were retrofitted with cooking stoves in the nineteenth century. As Germans started to emulate their English neighbors, they built end chimney houses with fireplaces or stoves in both the kitchen and the traditional stove room. As the years went on, many families added a new kitchen behind the old *küche*. The old room was converted into a combination dining room and family room. The old fireplace had often been fitted with doors that were closed in summertime when the great fireplace was not used. This was to keep the birds and animals out of the house. Later, the flues were sealed and the old fireplace often became a closet. In the twentieth century, many old central chimney stacks were torn out to give the house more floor space as demanded by modern living.

John Rasmussen (1828-1895) created "Berks County Farmscape," which is among the finest existing early views of a prosperous Pennsylvania German farm. Painted c.1879-1886, it emphasizes the neatness and order characteristic of the Dutch countryside. The house, once stone, has been enlarged in brick and is on axis with the well-maintained bank barn, which itself has been enlarged with an extra haymow on the left. An orchard is visible to the left. A neatly maintained vegetable garden is next to the house. In the field next to it hay is being loaded. A cow and several chickens are in the barnyard. Well-dressed men, with top hats, and a woman stand near the white-picket-fence-enclosed farmhouse. The farmstead faces onto a road, being traversed by both a fine carriage and a Conestoga wagon. As an extra bonus, the railroad borders the dirt road across from the idyllic scene. Note that the engine gives off only the faintest puff of white smoke. *Oil on zinc-plated tin 26 3/8 x 35 3/8 inches, Collection American Folk Art Museum, New York, Promised gift of Ralph Esmerian P1.2001.63*

As soon as they could afford to, many Germans built summer kitchens, separated from the house, where cooking and sometimes washing could be done away from the house in warm weather. First built with end wall fireplaces, many summer kitchens, like the *küche*, were retrofitted with stoves while later buildings were constructed only for stoves. It is interesting to note that in many households, frugal householders owned only one cook stove. An annual chore was to move the cook stove into the summer kitchen in summer, and back again in the fall. In some areas, stoves were not adopted for many years. The Reverend E. O. Steigerwalt recalled that his grandmother insisted on cooking over an open fire until her death in the 1920s.

Many times, summer houses were fitted with bells, often in a small belfry, which were rung to call workers from the fields for meals. Bells were also sometimes mounted on bell poles near the back door. The movie *Witness* filmed in Lancaster County in 1985, features a dramatic scene where an Amish boy in peril rings the farm bell repeatedly to bring help from the neighbors. This represents another use of the bell in the pre-telephone times; a usage still followed by the Amish. In 2005, as a twentieth anniversary event, you could take a tour of the summer kitchen where pivotal scenes of the Hollywood movie *Witness* were filmed.

The summer kitchen on the Landis brothers' farm is attached to the house – and it was enlarged at least once. On most farms the function of the summer kitchen has changed. Some have been attached to the main house by a corridor (some were attached before by an open arcade). Summer kitchens today are utility rooms, family rooms, home offices, or just places to store stuff. On many farms, the summer kitchen has simply been allowed to rot away.

The *stube*, or stove room, was the most desirable room in the house. When it was heated, it was warm and devoid of smoke and had fewer smells than the kitchen. Not all kitchen odors were desirable – remember the chickens in their cages. The stove room was often heated by a five-plate iron stove fed through the back of the kitchen fireplace or by a similarly fired masonry or tile stove.

The *stube* was a multipurpose room. It was a sitting room, but it was also used for home crafts and probably for sleeping, as well. In the late eighteenth century, emulating the English, the *stube* was transformed into a parlor. By the early nineteenth century, this was its usual function. The parlor became a special occasion sitting and receiving room. It is at this time that a second front door is commonly incorporated into the Pennsylvania German house, leading directly into the parlor. This separates the family entrance from the new company entrance and eliminates the need for a center hall.

Fire has always been feared on farms. One reason for building bake ovens as separate buildings was demonstrated when a display bake oven shed caught fire and burned. Fortunately the masonry was not destroyed and the structure could be rebuilt quickly. *Landis Valley Museum*

Separated from the back of the *stube* is the bedroom, or *kammer*, which was the most private and remote part of the house. The *kammer* was never heated. The *kammer's* function also changed over the years. While in conservative households it might continue as a downstairs bedroom, in others it might become a study or den, as at the Landis brothers' house at the Landis Valley Museum. In many instances, as the concept of the multipurpose "living room" replaced the Sunday-only-parlor, the wall between the two rooms was removed to create one large room.

If the house has a second floor, the enclosed staircase that ran along the outside wall, opposite the fireplace, would open into an open rectangular room onto which the chambers, or *kammeren* opened. Like the chambers below, these rooms lacked heat. The most desirable ones were those through which the chimneystack rose. These were the warmest of these upper story sleeping rooms.

Many houses like this still exist. When one of the author's colleagues and his wife were house hunting in 2003, they were amazed to be shown a house with "'gravity heat' on the second floor." The Ruizs were curious. "Gravity heat," they reported, consisted of iron grate-protected holes in the floor. "After all," the realtor told them, "Heat rises." Here, too, while some older houses are heated with stoves, some traditionally, and others because of rising energy costs, most have modern heating systems, often steam or hot water. Central air conditioning is also common. Outdoor condensers are a common sight, or blight, on the landscape.

When indoor plumbing was installed in many of these houses in the early twentieth century, it was common to have the toilet and the tub placed in the hall-room at the head of the stairs. A curtain might provide some privacy. In later years, a bedroom would be sacrificed to a bathroom and to provide closets. Often the space under the stairway on the first floor might be converted into a half bath. One of the most common additions to a traditional house is designed to provide extra bathrooms and closet space.

Many early houses were often one story topped by a high double attic which housed sleeping areas below and grain storage above, as at the Hans Herr House. On early farms, grain was too valuable to family survival to risk keeping it in the barn, and accordingly it was kept closer at hand.

Jacob Souder of Berks County stands next to his draw well or *Ziegbrunner*. Empty buckets lowered into a deep well on a rope attached to a crank mechanism were lifted filled with water. This device is simpler and less expensive than a pump. Strong muscles were required to draw the water. *Schwenkfelder Library and Heritage Center, H. Winslow Fegley, photographer*

Under layers of modern siding, many surprises lurk. Half timber, or *fachtwerk*, houses were known throughout Europe and the British Islands. This perhaps was the construction technique most widely used in our earliest German houses. In half timbering, which is the product of a wood-poor culture, the building is framed with timber. The areas between structural elements are in-filled with a variety of materials – often woven branches over lath, sometimes stone, or even brick. Characteristically, the infills are plastered over and left exposed, inside and out. Because of the constant maintenance needed by the plastering, our cold winters, and an abundance of inexpensive wood, most American half-timbered houses were sheathed over many years ago. As fashion changed, most interior walls were also covered over. Half-timbered structures, which are also almost always joined with pegged mortise and tenon joints, are extremely strong, and if foundations settle, they can display amazing flexibility. One Landis family homestead, which survived into the early twentieth century, was *fachtwerk*. Photographs of it as a decayed near-ruin survive. There are no complete *fachtwerk* buildings on the Landis Valley site today.

Log construction entered the American consciousness *via* the Palatine Germans coming here in numbers from the 1730s. The chimney placement, floor plan, and log work were distinctively Germanic. German log technology employs rough-hewn logs that often require very substantial chinking and plastering between them. A log, and they were always worked green, could be shaped by a skilled worker using a broad axe in about thirty minutes.

The groundcellar or *grundkeller* was used to store fruits and vegetables over the winter. A well-constructed one would maintain a year-round temperature of about 54 degrees. Most originally had earthen floors. Approximately 12 x 30 feet, this arched cellar was built under a portion of a 1730s stone tavern in Lititz. The current owners, who run their property as a bed and breakfast, are now taking up the concrete floor to restore the storage function of their "hidden treasure." *Lancaster New Era, Andrew P. Blackburn photographer*

Log architecture is very well represented at Landis Valley, but not in buildings original to the site. The most significant historic log structure is the Erisman House, which was moved to the museum from West Orange Street in Lancaster City in 1962. Built sometime before 1840 the building, in historic times, has been completely sheathed with clapboards on the front and less expensive vertical siding on the sides and back. The house is named for its last private owner, George Erisman, who operated it as a doll repair shop. A typical *flurküchehaus*, today it is set up as the shop and home of a successful early nineteenth century seamstress. Because, like most surviving log buildings, it is completely covered, few people recognize it as a log building. The same is true of the Print Shop, which is housed in a log building that once served as the parsonage for the still active Mennonite Church that adjoins the Landis Valley Museum. At one time it perhaps served as the meetinghouse itself. The side and back walls are log, but the front is *fachtwerk*. This is a very common continental feature seldom seen in surviving Pennsylvania architecture. One corner of the rear wall is covered in Plexiglas, rendering the logs visible to visitors.

The most obvious log structures are recreations: "The Log Farm," which was built in 1969-70 to represent a late eighteenth century Pennsylvania German farmstead. Such a farmstead would have been on a farm of about 135 acres with perhaps fifty cleared for farming, and it would have been capable of supporting a family of eight or ten or more. In addition to raising field crops – notably corn, wheat, rye, oats, spelt, flax, potatoes, and cabbage – the complex would have had a fenced raised-bed vegetable garden. The small house not only would house a large family but also was the center of food preparation and storage. Several recreated outbuildings are in the complex as well. The most important of these is the single-story log groundbarn or *grundscheier*.

An interpreter, dressed as an early nineteenth century farmer, walks the Log Farm. The house and ground barn are in the background. *Landis Valley Museum*

Germans are a very conservative people who strongly believe in building for the ages. Above all, Germans respect permanence. Accordingly, the most desirable of building materials is stone, which is also the most difficult material with which to work. The two basic forms of stonework are to employ ashlar, or carefully cut stone, or to use found or rough shaped, so called "rubble stone," for the walls. Beginning in the late eighteenth century, the Germans also started building in brick, a form of masonry easier to work than stone. Brick became the dominant material used by builders seeking permanence. Because brick walls have a tendency to spread, iron tie rods were often inserted through the house. Where they emerged, they were fastened with turnbuckles. Some of these were plain circular disks, others were stars or elongated "S" shapes.

A laden farm cart travels between a rye straw thatched hayrick or *barrick* on the left and a thatched ground barn at the Landis Valley Museum. *Landis Valley Museum*

Hogs were the preferred meat source for the Pennsylvania Germans. Sometimes the hogs were allowed to graze in the fields. More commonly, they were penned like these Tamworth and wild European boar crossed hogs at the Log Farm at the Landis Valley Museum. *Mr. & Mrs. Michael B. Emery Collection; Landis Valley Museum*

Stone construction was apparently especially nostalgic to George and Henry Landis. As children they had lived in masonry structures, but there were no extant stone buildings on their museum site. When the Oberlander Trust agreed to build exhibition buildings for the emerging museum, stone was chosen as the material. The Tavern, the courtyard sheds, and the gun shop were built in 1940 using local limestone. In later years, the Commonwealth built a stone-ended bank barn and a stone springhouse elsewhere on the site. Several of the most significant original buildings are brick, including the Elizabeth and Jacob Landis Farmhouse, which incorporates stonework from an earlier house, believed to have been on the site. The present structure was completed in 1828-29. Around the time of the couple's second son's marriage, about 1840, they built a small retirement house next to their original house into which Jacob and Elizabeth moved. After Jacob's death in 1848 Elizabeth lived on for many years in what came to be called the *Grossmutter* House, a form common on Dutch farms and still a common feature of Amish farms. The most impressive brick structure on the property is the Landis Valley House Hotel, which was built in 1855-56 for Jacob Landis, Jr. It became a social center for locals as well as a place for travelers. Henry H.'s diaries from the 1880s often mention it. He went there to meet and to drink.

Germans favored steep gable roofs. More elaborate houses might have gambrel roofs. Three roofing materials were traditionally employed: tile, wood shingle, and thatching. German roof tiles, formed and baked by tile makers, or *zieglers*, were made of the same red clay used for brick making and the celebrated Pennsylvania German redware pottery. Tiles would be hung in overlapping ranks on horizontal support elements called purlins. Roof construction had to be especially sturdy to support the heavy tiles. These weight-bearing structures could also easily support slate roofs, which became especially popular in the late nineteenth century with the development of the slate industry in Lehigh County. Several original tile roofs survive on houses and outbuildings in Lancaster and Berks Counties. At Landis Valley there are tile roofs on the recreated outdoor bake oven and the stone springhouse, outbuildings at the Log Farm. The Isaac Landis barn has a slate roof.

German style thatching differs significantly from the more familiar British Islands form where lengths of reeds are laboriously sewn together before being incorporated into the roof covering. Germans used rye straw tied with the same rye straw into small bundles called *fackles*. These, in turn, are tied onto a purlin roof to create a thick, weather tight roof. Rye straw, a byproduct of grain production, was favored because of its durability and its natural insect repellent qualities. Thatching on homes apparently died out by the mid-nineteenth century; however, thatching for agricultural structures continued into the early twentieth century. When first constructed, the barn at the Log Farm was thatched but, because of maintenance problems, it was redone in shingle. A hay barrack remains as an example of a thatched structure

Wooden shingles were the most widely used roofing material. German shingles began as shakes, eighteen to twenty-four inches long, split from log sections (often oak) and then tapered into shingles on a shaver's bench, *ein schniztlebank*, with a drawknife. Shingles would be overlapped side by side with a single nail holding each in place. Because of the length of the shingle, in fact, each nail penetrated and secured several shingles. An example of this type of roof can be seen on the log farmhouse at Landis Valley. Wood shingles, however, applied in the traditional manner have disappeared from non-museum settings because of the expense of the material and its fire hazard. A properly installed traditional shingle roof will often allow daylight in, but it does not allow rain in as it swells shut; however, snow can blow in. Descriptions of loft sleeping farm children who woke up covered in snow are common. Today, there is an increasing popularity in the use of wood shingles, which are treated with a fire retardant and installed in the more familiar Anglo-American tradition. These roofs are seen throughout the Landis Valley sites.

About the 1840s, standing tin roofs became very popular in German regions. These durable, and once inexpensive metal roofs, replaced many shingle, thatched, and tile roofs and were the roofing of choice for much new construction. Many of the early roof installers were tinsmiths whose traditional crafts, the making of household utensils, was threatened with the widespread availability of factory made goods. There is a tin roof on the Pierce Landis barn as well as on the new/old firehouse. Metal roofs, in turn, yielded to the cheaper still familiar asphalt roof shingles.

All traditional German houses, like their Anglo counterparts, had privies, or outhouses. In rural areas, these survived in numbers well into the second half of the twentieth century and are still used on occasional farms. Many other outhouses are used as tool sheds. The Pennsylvania Germans are traditionally reticent about sexual jokes, but an inordinate number of their jokes are scatological and especially outhouse centered. "Roadside America," a pioneer Pennsylvania German tourist attraction near Shartlesville, features a miniature train through Dutch countryside. Spectators can even see into an outhouse with a miniature Dutchman sitting on the board. At Landis Valley there are several outhouses, including a fine "for show" one at the Jacob Landis Homestead. Modern tourists, however, require modern facilities.

Early non-German travelers traversing areas of German settlement seldom commented on the houses or outhouses of the regions, but they were lyrical about the fertility of the soil and the magnificent barns which dwarfed dwellings. The Germans were among the first farmers to appreciate the value of regularly manuring their fields. To collect manure, you had to contain your cattle for a good part of the day, and so barns with animal byres became important. Manure was traditionally collected and piled between the house and the barn. German culture held: "A fat wife and a big manure pile show a man's rich." Ripening manure might also be banked against the north (and often windowless) wall of traditional stone or brick houses. The exothermic reaction of the decomposing manure would heat the cold northern wall of masonry houses.

While many early barns, like that at the Log Farm, were simply ground barns, i.e., one story and loft buildings where a central thrashing floor separated animal stalls and hay storage, the great classic Pennsylvania German barn is the bank barn. This is a two level and loft barn ideally built into a hillside, but often the second story is accessed by a ramp. The lower level contains the animal byre. The forebay of the upper level of the barn protects the farmer as he tends to his animals. The area under the south-facing forebay effectively catches low rays of the winter sun and keeps the area noticeably warmer than elsewhere in the enclosed barnyard.

Slaughtering and butchering hogs was an autumn ritual on most Pennsylvania German farms—and is still done on a few. It is an annual event at Harvest Days, held each October at the Landis Valley Museum. *Landis Valley Museum*

The Dryhouse or *Daerrhaus* at Landis Valley Museum was built in 1852 behind a still existing farmhouse on the Harrisburg Pike near Landisville, Lancaster County, by Levi G. Getz and his wife, Maria L. Landis, who had been born at Landis Valley. Last used during the Depression to process food for the poor, the structure was moved from its creek-side site to the Landis Valley Museum in 1975. Transported on a flatbed truck, other modern equipment was used to site it on the Isaac Landis farmstead at the museum. An interpreter shows its mesh-bottomed drawers – used in the drying process. A stove in the middle of the building provided the needed heat. *Landis Valley Museum*

Antecedents of the bank barn exist in the British Islands, as well as in many parts of Europe; however, in America the barn form is usually found only in areas of German influence. While there are many variations of the German barn, all of which are completely studied in Robert F. Ensminger's, *The Pennsylvania Barn: Its Origin, Evolution, and Destruction in North America* (1992), there are two basic forms of the barn. The most spectacular is often the "Sweitzer" or "Swiss" version which has a cantilevered forebay. The "German" or "Dutch" barn has a supported forebay. Often the supports can be very aesthetically pleasing. Additionally enhancing the beauty of a barn are ventilators. It is very important that grain and hay storage areas be well ventilated. Dust suspended in stagnant air can be explosive. The simplest ventilators are often simply slits. Nineteenth century mythmakers often said these were gun slits to protect against Indian raids. Later barns often have window size louvered openings and louvered cupolas on the roof, as well.

Barns are built of log, frame, or masonry. Prosperous multi-generational farms will often have an early ground barn, or *grundscheier*, now used for storage, and a semi-working bank barn and a collection of metal sheds. Framed barns continue the tradition of mortise, tenon, and peg construction of *fachtwerk*, or half timbering, but they are not infilled but simply sheathed. Because fire is an ever-present danger due to spontaneous combustion, dust explosions, or lightning strikes, many masonry barns are built with sturdy supporting walls. The remainder of the barn is of easier to replace materials. Early barns were usually small because they had only to house the few animals needed for subsistence farming. With the post railroad rise of commercial dairying, barns became ever larger and grander.

One of the Dutch country's best known and most loved barns is the landmark "Star Barn" in Dauphin County, which is in sight of a superhighway, Route 283, near Middletown. Built by William Motter (1822-1901) in the 1870s, it is clearly a gentleman's barn. A wide range of farm buildings, some of which remain, whose weathervanes, now gone, expressed each structure's use, originally surrounded it. The barn's great louvered gable-end stars and its Gothic lath-filled "windows" are ventilators designed to ensure that dust never built up to combustible levels in the "Cathedral of Barns." Saved from decay by the Millport Foundation, a local organization with Dutch roots, its future use is uncertain. Landis Valley has several bank barns. The best old ones are those on Isaac Landis and Pierce Landis homesteads. The Jacob Landis homestead's barn was demolished in the 1930s and its timbers were used by Henry and George Landis to build the Yellow Barn exhibition building. When the Pennsylvania Historical and Museum Commission wanted to construct a replica of a time-appropriate barn for the Jacob Landis farm, the original site was unavailable. Accordingly, the barn that is representative of a period Dutch bank barn is incorrectly sited: behind the house, rather than on axis.

Barn functions have changed, and today many are obsolete. With their large wooden surfaces in continual need of paint, barns are expensive to maintain. They have been described as "America's vanishing landmark." Many barns have simply decayed, others have been demolished for their lumber, and some have been carefully dismantled and moved to other parts of the country to be re-assembled, often as homes. Still others have been the victims of arson. Barn fires are spectacular and have commonly been set by arsonists – often thrill seekers.

The landscape is dotted with barns which have been adapted to other uses. The most fortunate are maintained on gentlemen's farms, often housing a few horses. Some have been turned into homes, like one visible from Route 441 in Bainbridge in Lancaster County. Others serve as

A woman drying "snitz": apples cut into eighths. Sweet dried apples, unpeeled, were used in cooking and making *Hutzelbrieh*, stewed dried fruit. Sour apples were peeled, dried, and used in pies where the combination of sweet and sour flavors was highly prized. The trays used for drying these apples were specially made to fit the outdoors baking oven used for fruit drying after the household baking was done. *Schwenkfelder Library and Heritage Center; H. Winslow Fegley, photographer*

garages on "farmettes," a term real estate folks have coined to describe a cluster of farm buildings that remain on a small plot of land after the farm acreage has been sold. Others serve as artist studios or are even used to house indoor swimming pools. The saddest ones, for the author, generally are those that have been unsympathetically adapted to commercial use. Although there are exceptions, a barn near Mount Joy, for example, has been turned into offices for Hess Associates so sensitively that you hardly notice the transition. More often, barns are converted into furniture stores or, especially on tourist routes, into "Dutch gift shoppes."

In addition to houses and bank barns, there were many other specialized structures that made up a farm complex. Not all farms had all of these outbuildings and many of them have disappeared. It is not at all unusual to see an old farmhouse, devoid of any of its ancillary structures, barn included. On the other hand, a number of these supporting buildings were made of masonry and have survived. Amos Long, Jr.'s *The Pennsylvania German Family Farm* is the encyclopedia of these specialized structures. We have already mentioned the cow barn, but there were other types as well, including hay barns and tobacco barns. Hay barns or *Holscheier* were found on very large farms where they would store hay and straw after the mows in the bank barn had been filled. Many of them replaced the hay barrack, which had been widely used since colonial days. Tobacco barns were used to dry the plants after harvest. Usually the walls were built with hinged vertical boards, which allowed air to freely circulate to expedite the drying process. As tobacco growing has declined in profitability, tobacco barns have been disappearing rapidly. Also used for storage were corncribs, *der Welshkanngripp* or *Welschkannhaus*, used to store husked dried corn on the cob. In addition to relying on basements and springhouses for food storage, many farms also had a cave, or a root or ground cellar, *der Grunkeller*, where foods could be kept over the winter. Earth-sheltered or excavated, these "hillside refrigerators" were prized in the winter because of their stable, cool climate, which was used for keeping apples edible as well as cabbage and a whole variety of other fruits and vegetables. This was also where sauerkraut would be stored. This was remembered by John H. Van Gordon, whose mother was Pennsylvania Dutch, in his book *Country Traditions*:

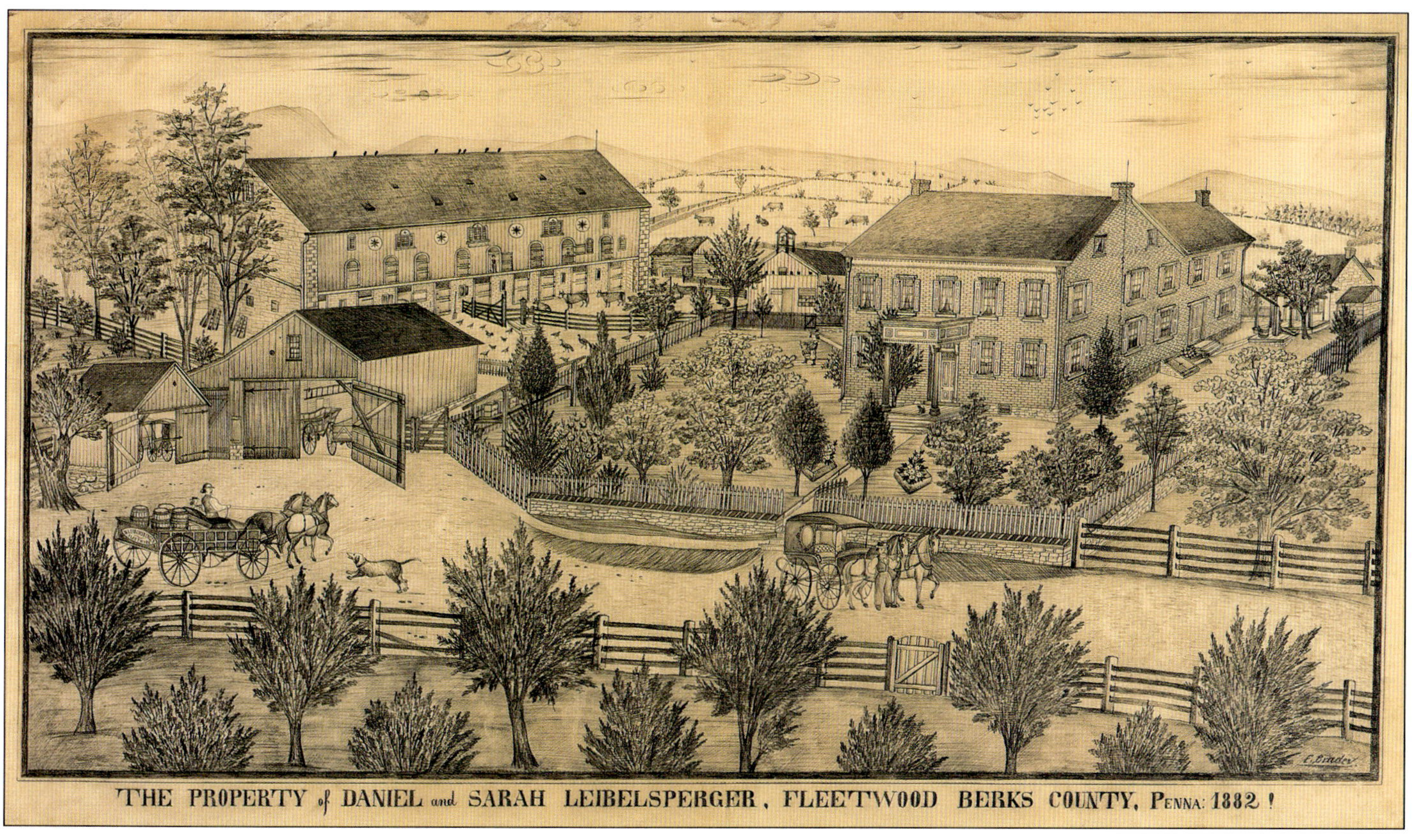

Ferdinand A. Brader (1833-after 1895) made his living drawing farmscapes and industrial landscapes in Berks, Lebanon, and Lancaster Counties in Pennsylvania and later in eastern Ohio and Indiana. "The Property of Daniel and Sarah Leibelsperger" near Fleetwood in Berks County is typical of his work. The brick English Georgian-style farmhouse has been enlarged many times. The lawn around the house is unusually large and elaborately landscaped. The large barn is decorated with hex signs. Of the numerous other buildings, most interesting are the open carriage shed (far left) and the adjoining open wagon shed. Appropriately, we see both a loaded wagon and a carriage on the road. *Pencil on paper, 30 5/8 x 51 3/16 inches, Collection American Folk Art Museum, New York, Promised gift of Ralph Esmerian, P2.2001.62*

> Every fall we made a big crock of sauerkraut. And when I say big, I mean about thirty gallons! We shaved the cabbage on the big slaw cutter, and after properly salting it, we'd pack it in the big crock. Father had cut a thick stone; it was cut round so that it would fit the crock and compress the cabbage. In a few weeks, we had very fine sauerkraut, enough to last all winter. When baked in the oven with pork chops, it was delicious.

Van Gordon also remembers how sturdy root crops were preserved out-of-doors in pits.

> When Mother thought it time to get them settled for the winter, she'd oversee us boys in burying the potatoes.
>
> Father would select a level, well-drained place. We would smooth it with a rake, hoeing up the edge, and making sort of a pie dish five or six feet in diameter. We then piled the potatoes in this spot, shaping the heap like a cone and covering them with six or eight inches of rye straw, which was placed so any moisture would run to the base. This mound was then covered with five or six inches of earth and the procedure was repeated, making a double thickness of straw and earth which was sufficient insulation for the most severe winters. When we removed the potatoes after the last frost in the spring, they would be about as fresh as when we placed them in the ground. We did the same for turnips; I believe this improved the turnips mightily. They got very sweet, and if you don't like them the way you get them now, children, you should bury a peck or two of them and see how wonderful they are when you dig them up!

Various animals, in addition to horses and cattle, had their own housing needs. So farms might have a sheepfold, *der Schoofschtall,* and it almost certainly had a pigpen or pigsty, *der Seischtall* or *der Seiben*. Sheep were raised mostly for their wool; pigs were raised as the preferred meat source. Chicken houses were virtually unknown on early Pennsylvania German farms as chickens and other fowl were on their own and allowed to roam freely. Fed only during the winter months, they were true free-range birds, most of whose diet was "worms, grubs, bugs, scattered seeds, and weeds." Today "free-range" birds are kept in large enclosures and grain fed. The hardy ancestral birds built nests wherever they wanted and children's work was to find the nests and collect eggs. When eggs became a profitable commodity, a chicken house or *Hinklehaus* was added to the farm. Farmwomen especially prized their "egg money" as their personal preserve. Emma Landis, for example, kept chickens and the eggs were exchanged for groceries and cash at the local store – or occasionally in Lancaster. Chickens for eating were customarily dispatched in the barnyard by the use of a sharp knife or axe, usually wielded by the farmwife. Slaughtering cattle or pigs was more complex and usually men's work. On early farms the butchering took place under the barn's forebay when the weather turned cold, usually in November or early December, with additional butchering being done in the springhouse or summer kitchen. The summer kitchen in the author's house still has floor drains used during butchering. In later years, many large farms built their own butcher houses or *Schlachthaus*. Many of these animal facilities were multi-functional or attached to other structures.

Closely related to meat butchering was the smokehouse or *Schmokehaus*, which was usually located very near the farmhouse and close to the outside privy, which it often resembled. In the absence of a smokehouse, meat might be smoked in the attic of a summer kitchen or a springhouse. Landis Valley's smokehouse is round and made of brick. Many others were made of stone or wood. The most famous Dutch apple variety, "The Smokehouse Apple," was discovered as a sport (a botanical term for a spontaneous mutation) growing behind a York County structure. Also important to the preservation of food was the dryhouse or *Daerrhaus* used to dry fruits and vegetables. *Snitz* (dried apple slices), for example, a diet staple, was used in winter pies and specialty dishes. Dryhouses were less common than the smokehouse but they were especially popular in fruit growing regions. On many farms the bake oven was used for drying, after baking was done. Dryhouses were generally small and had wire bottomed drawers or racks arranged around a central stove. Landis Valley's, a historic structure moved to the site, is one of the very few survivors of this building type.

The icehouse, *Eishaus*, was also very important for food preservation, especially of perishables like dairy products as a supplement to the springhouse. Constructed in a variety of shapes, some partially submerged in the earth, most were insulated with sawdust. In a properly constructed icehouse, ice harvested from a frozen pond in the winter could last until the next harvest season. Essential for helping to preserve dairy products, ice was treasured for cold drinks and the occasional making of ice cream – a special event that Henry H. Landis noted in his diary. Icehouses were either sited near the harvest site or, more commonly, near the milk house, or *Millichhaus*, which became necessary with the emergence of large-scale dairying after the Civil War. It was here that milk and cream were separated and butter and cheese made. All these functions continued to be done in springhouses or kitchens on smaller farms.

Rarely surviving on farms are the malthouse, *Malshaus* or *die Brauerei,* and the stillhouse or *Brennhaus*. Both became victim to changes in customs and laws. Brewhouses were among the earliest outbuildings constructed on farms. Homemade beer, along with

Henry H. Landis (center) and two workmen pose in a farm wagon probably used to haul limestone to the kiln he operated, which, no doubt, looked very much like this Berks County example. *Henry K. Landis, photographer, Landis Valley Museum Archives; Schwenkfelder Library and Heritage Center, H. Winslow Fegley, photographer*

hard cider made from apples grown on the farm, were the principle beverages consumed by farm families from pioneer times until well into the late nineteenth century. Stills came a little later, but by 1800 one scholar claims, "... there were more distilleries in Pennsylvania than gristmills" and "From 1810 to 1840, nearly one fifth of the farmers of York County owned a 'copper still' by which they distilled their own cereals into whiskey." Whiskey was not only almost universally drunk, but it was efficient to convert grain into this easily transported and profitable product. Landis Valley has a grand copper still in its collections. Wine was also made on farms, often in small quantities, often by the housewife, but in areas where grapes were abundant there were small wineries and wine cellars. On smaller farms the wine press was also used for cider making. In addition to grapes, wine was made from many other, occasionally unlikely, natural products including cherries, blackberries, elderberries, mulberries, strawberries, plums, rhubarb stalks, rose petals, and the flower of both white clover and dandelion. In his later years, Henry H. was almost obsessive in recording the numbers of dandelion blooms he could pick ... often thousands on a spring day.

Other structures served many other functions. In addition to the springhouse, there were other water-related buildings, especially the pump house or *Bumphaus*, for those who were unfortunate enough not to have a spring. This might be a simple shelter or an enclosed, occasionally elaborate, building. By the eighteenth century, however, some houses actually had indoor wells and pumps in the kitchen or *küche*: a precursor of indoor plumbing. Near the water source, in later years, there would often be a washhouse, *Weschhaus*, supplementing the summer kitchen. In mild weather laundry would be done outside, but in cold weather the washhouse was an important structure. Most had a fireplace or a stove for boiling the water needed for clothes washing.

Wood fires (and later coal) were used for heat and cooking everywhere on the farm. To store firewood, virtually every farm had a woodshed, variously called *der Holsheisel*, *der Holsshup*, or *der Holsblotz*. Many were open sheds, but others were closed buildings, usually located near the kitchen, summer kitchen, and washhouse.

Many farms also had a wagon shed, *der Waageshupp*, often a large building to house the increasing numbers of wagons and carts used on the farms as the years went on. More prosperous families might own an enclosed carriage and/or a buggy like the Landises, and, if so, the wagon shed would serve double duty as a carriage house. To keep the property in repair many farms also had a tool shed, sometimes combined with a small smithy, *der Gschaerr un Schmittschopp*. Both varied in size and they enabled a farmer to keep his tools in good order and to make minor repairs.

Lime was very important to agriculture and numerous farms had their own limekiln, *der Kallickoffe*, built of stone. Some farmers, like Henry H. Landis, made a business of selling lime to his neighbors. The site of Henry's kiln is buried under a suburban development across the road from the Landis Valley Museum on land his sons sold many years ago.

Every farm community needed its grist mill, used to convert grain into flour. Often built of stone, many of these early water-powered mills were adapted to steam and later diesel power, and many operated into the early twentieth century. A few limped on even longer. Some are run as tourist attractions today. Many are derelict; others have been converted into antiques shops or homes. They remain a common sight in rural Dutch country. Benjamin Landis had interests in mills and the family married into the Binkleys of nearby Binkley Mill as well.

Dutchland towns are distinctive and often blur a distinction between farm and village. The smaller communities tend to be linear, often one street deep, with the houses built right up to the property (now the sidewalk) line with no front yards. In many towns, like Stouchsburg in Berks County, the deep backyards end in small barns used for keeping a cow or two for family use. Backyards certainly housed chickens, perhaps a pig or two, and caged rabbits, as well as a vegetable garden. German settled towns tend to end in "burg" as in Strausburg, or the even more Germanic "dorf," as in Wormelsdorf. Towns settled by British Island folks often end in "ville," like Wellsville, but many of these soon had predominantly Germanic populations, as well.

Of the cities in the region, Lancaster perhaps retains the most German atmosphere, although York and Reading also have some strong Germanic overtones as well. Also bridging town or city and country, and farmers and consumers, are the very Germanic markets. Still surviving in the cities and in small towns in the countryside are market areas, sometimes called "market squares," occupied first by open market stalls, then, in some cases, by market houses.

Today, Reading, Harrisburg, Lancaster, and York have flourishing market houses, sometimes called "Farmer's Markets" by outsiders. Locals call them simply "Market," except in places like Lancaster, which had several market houses into the twentieth century. Only "Central Market" survives. Historically, markets were held on Tuesday and Friday. Now many have added Saturday hours as well. In the countryside, there are sprawling rural "Farmer's Markets," including the Green Dragon and Roots, both in Lancaster County. Both are twentieth century in origin, and both on their days of operations (Roots, for example, is only open on Tuesday) are wonderful places to see Plain people, and church people, and every other type of person intermingle in a fair-like atmosphere. Occasionally the Landises "attended market" (i.e. sold there).

A close examination of Henry H. Landis' diary, beginning in the 1870s and continuing into 1914, a dozen years before his death, provides an insight into agriculture at Landis Valley at its most intensive and usually at its most profitable. It certainly reflects many of the trends in the Manheim township of its times as studied so effectively in Steven S. Miller's comprehensive unpublished master's production, "A Finer Focus of Pleasurable Living," which is subtitled "Changes in Manheim Township, Lancaster County, Pennsylvania Agriculture 1880-1920."

Grain mills were necessary in any agricultural area and were commonly seen throughout the region. Early in the twentieth century, Henry K. Landis photographed what is probably the Hershey Mill near Mount Joy in Lancaster County. His father's (Henry H.'s) diary has numerous mentions of taking wheat and corn to mills; sometimes the grain was to be ground into flour, but mostly to be sold as a regular part of the family's income. *Landis Valley Museum Archives*

Removing stumps and clearing land was arduous work using hand tools, and even animal power. These young boys, residents of the Bethany Orphans Home in Womelsdorf, run by the Reformed Church, deserve their "Resting Time." Hard physical work, and especially farm labor, was expected of the rural poor. *Mr. and Mrs. Michael B. Emery Collection*

Work was a constant on the farm, but on this prosperous farm it was not always intense, nor was it always performed by the Landis family themselves. Henry regularly hired men as day laborers to help at his limekiln as well as at the farm itself – and they were often employed throughout much of the year, not just at harvest time. The names most commonly found in his account books are members of the Whitcraft, Lawrence, and Christ families. From the account books we learn that he often made cash advances to these men, sometimes for necessities like the purchase of shoes and other times for money was to go to the fair or to a party.

> "John Lawrence to Scythe & med I bought for him in
> Lancaster 1 35 [June 1878]"
> William Whitcraft
> to cash to go to Roseville
> to the party in the evening 1 00
> [Oct 1878]

The Franklin Pierce McQueney story is more typical of a hoped for long lasting relationship that failed, lasting, as it did, only from a February into March in the1880s.

> [February] 18
> Franklin Peirce
> McQueney began
> service for a term
> of nine month at
> twelve Dollars pr
> month
>
> March 21
> Franklin Peirce McQueney
> Quit service on the twenty
> First of March I paid
> Him seven dollars for the
> Time he was with me
> It was so agreed between us 2 00

Henry also used frankly itinerant labor. On Friday, August 11, 1876, he notes, "today my Gypsy man left after supper." At other times even his most regular help gave him problems as in May 1878, "Old Christ went to town this afternoon and came home drunk. I just left him alone and he staid in the barn overnight." But ownership had its perks. On a ninety-degree day in May of 1889, Henry records, "Billy and Omer are working at the fence Jacob is plowing and George and I are loafing."

Farming was, and still is, a dangerous occupation, especially since sharp tools and machinery are central to food production. Twenty-first century Lancaster newspaper accounts often have stories about deaths due to tractor rollovers or people being injured by getting caught in a machine. Pre-industrial accidents were also common and could be devastating. Henry H. recorded many in his diary.

> Thursday 26 [1883] this morning I started to cut Timothy seed [a hay related crop] along the fence about nine o clock I cut my finger I reached my hand through the fence for George to tie up my finger in so doing I dropped the sickle the point of it entered my foot Just below the instep and gave me a great deal of pain the blood flowed freely and now (3.o.clock) I am scarcely able to walk cold water seems to give me relief at once Billy is working in my Tobacco Ladd went off somewhere George is about, this eavening my foot is very painful and very much swollen

Henry survived to return to work the next day. In subsequent years he would endure back injury and trauma to an eye.

The household was Emma's, Henry's wife's, domain and there are frequent references in the diary to her cleaning house, cooking, tending the children and her chickens, doing the laundry, and myriad other housewifely tasks. But there was also help, especially with the arduous task that was nineteenth century laundry. Emma's only daughter, Nettie May, was born late to her parents and, while she did some chores around the farm and the house, she was never a drudge. She was clearly a cherished delight. "Hired Girls," as they were called, were a constant as were complaints about the girls, who were always leaving or not performing their chores as Emma would have liked. Sometimes the young woman would become sick. On Friday May 18, 1876, Henry notes that, "Dr. Bowman Came here this morning to see Sarah our girl she has Rheumatism and is confined to bed."

The employment history of one Sapora Wise is interesting. She was not only a hired girl, but she was also a tobacco sharecropper, a dressmaker, and a general errand runner. Her history with the Landis family was brief, from April to June 1872. She apparently even was paid a signing bonus, as seen in Henry H.'s account book

April 15th 1872
Sapora Wise began
service at Six Dollars
a month for a
period of two months
and after the expiration
of two month she is
to receive five dollars
pr month for the
balance of the year
She (Sapora Wise)
has also agreed to
farm a certain patch
of tobacco on the shares
She is allowed time
to attend to said tobacco
————May 16————
Paid Sapora Wise 3 5
————20————
twelve Dollars for
a new dress she bot
in Lancaster 12 00
————June 1————
Paid Sapora Wise
for making a dress
at frankforts 3 25
————June 18th————
Paid Sapora Wise
for a pair of gloves 40
————June16————
Sopora Wise quit
work on the 16th
of June she went home
to her sister on act of
not being well

Henry recorded a post script:

————Oct 3rd————
Sapora Wise paid me
Seven Dollar in
money she had over
drawn

While Emma may have done some sewing of clothing for the family, it is also clear from the account book that many of their garments were sewn by others or store bought as befitted a prosperous farm family. Emma was a lady of means who closely guarded her money, which was always accounted for separately. On the settlement of her parents' estate Emma inherited a substantial amount. "Emma get $3000 together with what she already has making all $5500 and her outfit when she left home which was rather liberal."

As in many marriages, farm-based or not, money was always an issue, with Henry believing that Emma was spending too much. On Thursday, September 29, 1870, Henry notes, "I am down in the Dumps [*sic*] about as deep as ever now I am hauling wheat and seeding at the same time and doing it all for some one not me the money is all spent before I get it in my hands my wife does the managing."

Like many farmers, Henry was often strapped for cash, and was often in debt to the local banks, but his assets were substantial. In December 23, 1871, Henry Harrison prepared an "Inventory of my effects on commencing [*sic*] or carry on [*sic*] on my business." It is revealing not only about his financial worth, but it also tells us a great deal about his farming operation.

one farm in Manheim Twp. Containing 96 Acres Valued @	19200 00
Small Farm Containing 12 1/2 Acres Valued @	7000 00
23 Shares of Manheim & Lancaster Turnpike Stock Valued @ $18-pr share	414 00
6 Shares of Turnpike Stock of the Lancaster & Fruitville Turnpike Co Valued @ $50—pr share	300 00
2 shares in the farmers western marker of Lancaster City Valued @ $50 valued	100 00
13 Acres of timber Land in Martic Twp Valued 20 pr Acres	260 00
1/6 interest in the Mount Union Iron Co Valued @	29500 00
2 horses Valued @	200 00
1 colt	50 00
2 cows "	100 00
1 two horse wagon Valued @	40 00
1 carriage Valued @	75 00
1 buggy "	40 00
4 head of fat hogs Valued @	50 00
20 chickens Valued @	5 00
300 bu of corn Valued @	150 00
100 bu of wheat "	150 00
100 bu of oats "	40 00
lot of tobacco Valued @	300 00
" of Hay and fodder Val@	25 00
500 Chestnut Railes Valued@	75 00
100 " Posts Valued@	30 00
100 Pine Railes Valued @	19 00
1 1/2 ton of coal cost .	9 30
6 cord of wood Valued @	24 00
lot of plank " @	16 00
5 shares in the Lancaster	

Building Association	
Valued @	75 00
Note Due from David Decker	50 00
Note Due from Michael Barch	39 00
Note Due from Daniel Sener	15 00
Amt Due from David Graybill	250 00
Amt Due from Henry Decker	100 00
Amt Due from I S Landis	25 00
Amt Due from Henry S Landis	40 50
Amt Due from Lancaster & Man=	
Heim Turnpike Co	105 00
Jacob Immel note	15 00
Sleigh Bells & harness	75 00
Cash on hand	30 00
Sundries	15 50
1/2 share in Engle Keller	
& Co Manufacturers of the	
Buckeye Dryer and Feed	
Steamer	250 00
Installment on same	47 50
1/2 share in the New York	
Co for the Sale of the Buck	
Eye Dryer and feed steamer	250 00
Installment of same	7 00
Conestoga patent fence	150 00
	59705 80

Henry's liabilities totaled 13,900.00 with a significant amount owed to relatives including his and Emma's fathers.

Liabilities	
H L Landis Due Apr 72	3000 00
Geo Diller Due April 72	2500 00
David Graybill April 72	1000 00
Henry Landis sr April 72	500 00
A H Kauffman Due Jan 721	115 00
A H Bomberger Due Jan 1872	100 00
H L Breckbill Due Jan 1872	30 00
E H Kauffman Due	6 00
Shultz & Bro Due	20 00
Socher & Son	100 00
Bair Q Shenk	300 00
Mechanics Bank	1100 00
Lancaster Nat Bank	2600 00
	11 371 00
Dr A K Rohrer Due Apr 1872	10900 00
Note Due Jan 3rd 1872	500 00
" " April 1st 1872	2500 00
	13 900

Notations in pencil noted when each of these cash debts was paid.

Ensuing years were not always kind to Henry. His largest single asset of 1/6 interest in the Mount Union Iron was rendered worthless, due to bankruptcy. To help offset his very substantial losses here, Henry went into the lime burning business and tried many other money making schemes, including selling the new phosphate fertilizer for the "Allentown Fertilizing Co" [1889], but sales were never brisk, although Henry felt that chemical fertilizers were often better for crops than manure. He never made the big payoff he craved.

Animals were important to their farm. Horses were used as draft animals and to pull their "buggy" and their "carriage," in modern terms, their Honda and their Cadillac. The Landises kept a few cows, mostly for butter making. In later years they stopped dairying entirely. By 1914, one of Henry's chores was to stop by a neighbor's farm for milk. They also would raise occasional steers for slaughter. The family also enjoyed veal, which they purchased from a neighbor. "Old Mr. Graves," Henry noted on April 28, 1876, "brought us a quarter of Veal it weighed 25 lbs nice and fat." Hogs were a constant crop, raised for their own consumption and for sale. Among the interesting insights in Henry's diary was his note, "I paid Elim Landis for a pig and the keeping of my sow and the service of his Boar $.25". Butchering a hog or hogs was an annual event on a cold day in November or December. At Harvest Days, celebrated at the Landis Valley Museum each October, a hog is slaughtered, butchered, and made into many products, employing every part of the pig "except the oink." Henry from time to time dealt in horses, steers, and hogs, which he regularly bought at Hilton's auctions, held in the village of Landis Valley. These were usually modestly profitable. He also made additional money pasturing horses for the auctioneer before sales. "Sunday October 16, 1889, Hilton Brought 60 head horses to pasture in the Field behind the Barn."

Chickens became ever more important as a cash crop for Emma as the years went by – as a source of eggs. Henry often reports trips to the store in later years where he would buy groceries and candy at the local store in exchange for eggs and some cash. At times they also raised guinea fowl, ducks, geese, and occasionally turkeys.

Field crops were clearly more important. Henry's various inventories point up the value of hay, fodder, corn, wheat, oats and potatoes but at various times other crops, including rye and clover, were also raised. Crop fields had to be fenced to protect them from stray farm animals and fence rails and posts were constantly in inventory and fences were constantly being worked on. Henry often talks about hired men mending fences. On May 19, 1876, for example, he notes, "Jacob Lawrence and his son Henry were here working at making fence to day [*sic*]." There was also an apple orchard and the Landis farm produced cider and apple butter.

Meadowland and hayfields were basic to a nineteenth century farm. In many areas some were even irrigated as early as the eighteenth century. Archaeology discovered remnants of the system that may have been constructed by the Boones at what is now the Daniel Boone Homestead, a museum property in Berks County. There do not seem to have been irrigated meadows at Landis Valley. Meadows, however, provided food for animals and pasturage, and for farmers near a large town like Lancaster, a source of both cash and manure. Henry's diary is peppered with references of his men regularly taking loads of hay and straw into town and returning with manure. Meadows, even those not irrigated, required work to maintain them, including re-seeding on occasion, which was done in either fall or spring. White clover was popular for meadows and was regularly seeded in quantity – not an inexpensive undertaking. In April 1881, Henry bought 5 1/4 bushels of clover seed from the nearby Nolts mill for $28.25. Different kinds of hay were also planted for harvest, including a European import called "Hungarian," which became popular in the later part of the century. Hay had to be "made when the sun shinned." If it got wet it would be ruined – and occasionally it did. "… 9 loads of hay is spoilt," Henry noted July 5, 1889, "for having to [*sic*] many rains." Hay had to be cut – and at Landis Valley this was done both with machinery and by hand. Saturday June 25, 1881, "Steigleman is cutting with the mower & John Goist Ben Bahuar are at it with scythe." After cutting the hay had to be turned to promote drying "George took three men and went to the hay field to stir up hay", before it was raked up and stored in hay mows in the barn or as stacks, often under a roof. "George and I went out … to fix the roof on the hay stack." Saturday July 8, 1876, Henry wrote, "to day we finished taking home hay last night we took home our last rakings and this morning we finished our stack we have one more full [*sic*] of hay we finished at supper and were all tired out and we did not work after supper." In later years the hay would be baled by a hay baler and stacked in the barn with the aid of an elevator apparatus.

Corn and wheat were the two most important grain crops, far surpassing oats and rye. Corn was mainly used for animal feed, mostly by the Landis family, but some was also sold. Corn was planted on plowed earth using a seed drill or planter. Like all nineteenth century agricultural products, the process from seed (which was often saved from the preceding year) to the final process was laborious. Fields of young corn needed to be cultivated to keep weeds down. Mature ears had to be picked, stalks cut and stacked into shocks to dry in the fields. Corn stalks would be turned into "chop" by simply chopping them into several inch-long pieces, and used as cattle feed. Dried cornhusks were used for human and animal bedding and as a precursor of toilet paper in the outhouses. Corn, of course, had to be husked first, which was hard work. Henry usually had hired help do this. On Saturday Oct. 8, 1870, "Billy husked corn and John hauled home two load our corn is very good large." In November we learn that one "Henry Greenly and John Laurence were each paid $23.46 for husking corn." Corn on the cob would be stored in corncribs until needed.

The meadow or *Schwann* or *Wiss* was the most important source of hay on early farms. Farmers favored well-watered meadows where a stream could provide water for irrigation as needed. Grass was cut at least twice a year with a scythe, hand-raked, and stored in the barn. To protect the meadow from wagon ruts and horse damage, many farmers hauled the grass on a wheelbarrow and then loaded it on a wagon for the longer trek to a barn. After the final cutting or the *Ohmet* or *Ulmet*, cows would be allowed to graze. *Schwenkfelder Library and Heritage Center, H. Winslow Fegley, photographer.*

Cows grazing in fields are still a common sight in Dutch Country. A herd grazes in the meadow in front of the Sheetz House, which is now the visitor center for the Conrad Weiser Homestead in Berks County. Hershey cows, in Derry Township, Dauphin County, provided milk for tons of the famous chocolate. Note the bank barn in the background, which was built in the early twentieth century. *Mr. and Mrs. Michael B. Emery Collection; The Cottage Collection*

Corn for chicken feed and making of meal had to go through a corn sheller – a great labor saver – that separated corn from cobs. Corncobs were used as kindling and fuel in stoves before a hotter fire was built with wood or coal.

Wheat, oats, and rye as crops were all handled similarly. Seed was broadcast or seed-drilled on plowed fields in fall or spring, depending on varieties grown. Henry often rented out his drill "Daniel Graver (a neighbor) had our Grain Drill in use to day" (October 8, 1889) The mature grain had to be cut using either a horse-drawn reaper that Henry owned or with scythes — always backbreaking labor. During harvest time neighbors who had reapers would often help harvest one another's crop. Henry also rented his reaper to less affluent farmers. After the grain was cut it had to be bound into bundles and stacked into shocks. Grain bundles were neatly and quickly tied with a strand of grain straw – often by hired help. "Billy is helping ... to bind wheat today" (July 13, 1883)

In July of 1889 the wheat harvest began Monday the 15th " ... we began cutting grain with five men at binding and Geo & Ladd at Shockings I carry water for the men everything works first rate." The crew worked feverishly for three days. On Thursday Henry notes, "Still hauling home wheat we made two Stacks in the field behind the Barn. This evening I am entirely out done." Youth had more energy: "George went to a picnic ...."

When the grain was dry it had to be separated from the chaff. First this was done by flailing or tramping the wheat. "Christ and I tramped out a floor of wheat" Henry recorded in 1876. By 1880 the grain would be threshed using Henry's horse-powered thresher. Later a professional team using steam-powered equipment would do the job. Now the grain would be ready for use. The straw would be used for animal bedding – often sold in Lancaster or exchanged for manure. Rye straw was especially valued because it repelled various insects. Rye, oats, and wheat, along with hay, always had to be harvested at around the same time ...usually during July, which was especially hectic.

Most rye and oats were used on the farm as was most corn, but wheat was an important cash crop. An old adage holds "that a farmer's footprints are the best fertilizer" and Henry believed in walking the farm on Sundays. On a pleasant May day in 1889, the farmer noted with pride that, " ... the wheat looks splendid." On Sunday July 19, 1914, Henry " ... took the two Dogs and walked out to the wheat field and examined wheat shooks [*sic*] to see if they were dry enough to "Thras [*sic*] tomorrow I found them in fair condition" and, indeed the next day Binkley and " ... his Thrashing outfit ..." showed up on the farm and " ... they Thrashed about 400 bushel of wheat and bailed straw ..." Wheat was generally sold to one of the local mills. On March 14, 1876, he notes selling wheat at Hostetters mill for "1.35 pr bu". Thursday July 19, 1883, "Billy hauled wheat to the mill today worked for me the whole day ...." Wheat cash was especially welcome.

Potatoes were also an important field or secondary garden crop. The tubers were planted in April or May and harvested from August through early October. Preparing potatoes for planting was time-consuming. Especially fine potatoes stored from last year or purchased would be moistened to encourage them to sprout. From time to time Henry bought seed potatoes from others – for new varieties. Homegrown or purchased, the tubers would be cut, making sure that each sprouted eye had a piece of potato attached to provide nourishment. Henry noted " ... the women cut nearly all the potatoes" May 2, 1889, and he and Jacob did the planting in rows enriched with phosphate fertilizer. As the potatoes grew the plants were susceptible to many insects, which Henry controlled by regular application of "Paris Green," an arsenic compound. When ready to harvest, after the tops died down, the potatoes would be plowed up with a special device, allowed to dry and then gathered by hand into baskets and stored in the root cellar. In September 1876, Henry records "Took up Potatoes today."

The favored varieties of potato that Henry planted were the Blue Peerless and the Early Rose. He especially favored the Early Rose, a variety introduced in 1870 that was "One of the most highly advertised potatoes of its time" and indicates that Henry was, as a progressive farmer, growing new varieties whenever possible. "Adam Landis was here to day helping to plant potatoes and spreading lime" (Monday April 24, 1876.) Potatoes grew best on well-limed soil.

Some potatoes were sold, but most were eaten on the farm by humans and animals. Henry tells us that he cooked potatoes before feeding them to his cattle. One potato sale was especially interesting. "This morning the first done Emma & Jacob Measured up 4 1/2 bu of potatoes I sold to H. L. Landis yesterday he is sending them out to the JohnsTown suffers [survivors of the Johnstown Flood] he is sending out 100 bu he pays us 25¢ per bu".

Every well run farm had its fruit orchard – especially apples. Wednesday April 26, 1876. "Hen and I," Henry recorded, "were planting apple trees until dinner ...." They also had a peach orchard, but no pear trees. The diary notes several times they bought these fruits. Apples were, of course, mostly made into cider. The Landises did not have their own press, but used a neighbor's. On August 29, 1889, "Myself & Nettie went to Haucks with apples and made 3/4 Barrel of cider brought it home put it over the fire and boiled apple Butter all done about sundown." In the Landis orchards, grass that could be harvested as hay was grown between the trees. The worst pests that they had to deal with in raising apples were tent caterpillars, which Henry destroyed using "... a torch and Burn them," ... a very effective tactic used today by the author.

It is also interesting to note crops that Henry didn't grow – and these were fibers, flax for linen and hemp. Both produced valuable seed as well as fibers. His ancestors raised both. His great grandfather Benjamin Landis' will probated in 1787 directed that "My sons Benjamin Landis and Henry Landis shall yearly and every year deliver 30 pounds of good hachelt [combed] flax or hemp unto my beloved wife Ann Landis." Another Henry Landis, also of Manheim Township, Lancaster County, directed in documents drawn up between 1834 and 1839 that his widow Ann (less generously) be given, " … my yarn, thread, hemp, flax and tow at my decease … [and] I do order and direct that my said son Jacob Landis and his heirs and assigns shall give and deliver to her at her dwelling yearly … of the lands hereinafter devised to him … 6 pounds of good hackled hemp or flax."

The rise of factory processed textiles had, by Henry's day, made the growing and processing of these home textile fibers unprofitable. Both continued to be grown sporadically in other areas of Pennsylvania for many years.

While flax processing implements were very familiar to museum founders George and Henry K. Landis, hemp had become so distant that in 1928, when they were exploring the ruined Fahnestock Mill near Millway, they "… found nearly half buried in the soil, a conical, hollow, millstone of sandstone and they asked, 'What is that?'" It was only years later that they learned it was a specialized stone used in processing hemp. Even before then they moved the mysterious artifact "Weighing probably two-thirds of a ton this stone required a chain block to load it on a heavy truck … for transportation to the museum." Where it remains today.

Of all the crops Henry grew, however, tobacco was the king and his favorite. Farmers often called it "the mortgage burner" because it could be so profitable to grow. Tobacco as a crop was both a blessing and a curse to Henry who made a lot of money raising tobacco as a farmer, as a renter of tobacco patches, and as the partner of sharecroppers. At various times he fancied himself as a tobacco merchant, buying and selling his neighbors' crops – often speculating – and often losing money. As a businessman, Henry was cursed with the opposite of the Midas touch. Everything he touched, commercially, turned into manure.

Tobacco was a mania for the family. George, Lad (son Henry), Emma, and even Nettie May had their own tobacco patches. Henry, always the doting father, often complained about having to do the work on Lad's patch occasionally – and on Nettie's always. It is interesting to note as you read Henry's diary that tobacco is almost always spelled with a capital 'T'… the only crop so honored. Tobacco was and remains a year-round very labor-intensive crop. Today the most successful raisers tend to be the Amish, whose large families provide the needed labor.

A woman, possibly Emma Landis, feeds the chickens at Landis Valley. As the years went on, Emma and Henry H. Landis paid increasing attention to poultry. The Landis farm had a chicken house or *Hinklehaus* and eggs were a continuing source of ready cash, often used for purchases of store-bought foodstuffs. *Henry K. Landis, photographer, Landis Valley Museum Archives*

It was a prosperous farm that had six, and possibly more, haystacks. On Sunday you could enjoy the fruits of your labors. These young ladies pose behind a rail fence. The young man on top the first stack probably helped with the harvest on this Hamburg area farm. *Mr. and Mrs. Michael B. Emery Collection*

Tobacco had to be started from seed in protected and very well cared for beds... this in late March or early April. "Geo Started Frank Burkhart getting Tobacco plant Beds ready for Sowing the seed." Young tobacco plants were vulnerable to all kinds of weed infestation, insect and disease problems; to counter this a new process emerged in the twentieth century. On April 2, 1914, Henry wrote, "this afternoon Abe Binkley began cooking the Tobacco beds I went down to see the thing done the ground got hot you could not put your hand in it after it was done we examined it found worms rosted George came home at 3. P.M. he also went down to see the thing done". Father and son were so impressed that two days later:

> George sent Nathan out to Martin Espenshade for Some pans to Steam his Tobacco plant beds It is something out of the ordinary it is done to kill insects and destroy weeds the above was put down by mistake for Thursday Rose at 6 Six took breakfast helped George off to Lancaster then I went down to the farm looked at the working of Steaming the Tobacco beds they finished at noon

In late May the plants would be transplanted into the fields. Literally thousands of plants were handled. Horses or mules were used to pull a specialized rig called "a tobacco planter". One or two farm hands sat close to the ground placing the plants, which had been harvested from the starting beds. Once planted, the tobacco required constant care. Friday, May 29, 1914, Henry notes, "this after our men were planting Tobacco they put in ten thousand plants they look well this evening". An animal powered water tank followed the transplanters.

Each patch or field had to be hoed and weeded. "I was weeding my Tobacco plants to day they do not look so poor since they are cleaned." [Henry's diary May 30, 1888] Farmers had to be alert for tobacco worms, which could be very destructive. Children were often paid to pick worms. Tobacco also had to be suckered (its side shoots removed). Henry's diary is peppered with accounts "suckered some Tobacco." Another chore was "topping," preventing the plant from flowering. "Topped 350 Stalks of Tobacco", "I toped four rows of Tobacco", were commonplace. Topping was done in late July and August.

Late August or early September tobacco was harvested. Each stalk was cut by hand with a specialized long-handled shear. Next the stalks were threaded onto lathes, which would rest on a special frame inset in wagons, which kept the hanging plants suspended. It was important that the leaves not be damaged because that would hurt the value of the tobacco. Driven from the fields, the tobacco would be transferred to a special barn with slatted or louvered sides where it would air dry. Henry's first tobacco barn blew down in a gale in July 1883 and by next month a new building was raised in time for the new crop. How do you celebrate your birthday if you are besotted with tobacco? Henry's diary for September 14, 1888, records, "my Birthday at home worked in my Tobacco".

Nettie May Landis watches a hired man do the fall plowing at Landis Valley. Note the huge haystack complete with a ladder. In front of her is the new tobacco shed (now demolished) built in the 1870s. Opportunistic chickens scratch for grubs and insects in the newly turned soil. *Henry K. Landis, photographer, Landis Valley Museum Archives*

The tobacco would cure in the barn until January when the next process was begun … "stripping" in which the leaves were stripped from the stalk, then sorted by size and quality, and finally baled. Friday, January 4, 1889, "Jacob and I [Henry] Striped Tobacco the whole day 140th lath carried 228 down in the cellar". The work was often done in the barn basement, heated by a stove.

The best time of the year was when the tobacco was sold. This would usually be in March or April. Weather and ensuing quality and demand determined the price. Sometimes there was great joy. At other times it was more restrained.

> March 31 — Thursday 1881
>
> Went to Lancaster this morning with two load of Tobacco sold to Lockenbrook & Co Billy Whitcraft and the lot the Boys raised the Boys raised about 4000 pla[n]ts and realized only 124.30/100 for theirs Whitcraft had about 10000 plants and realized only 201. 11/100 for his and mine shares the weather was awful disagreeable this morning up to noon snow fell fast the whole fournoo we came home soon after dinner I put in Bank $225. and checked out $100. towards eavening it began to snow again

Next year would always be better. Farmers have to be the eternal optimists and the ultimate realists. Pennsylvania German farmers, including Henry, have demonstrated these qualities for hundreds of years. Today there are no tobacco barns at Landis Valley and they are vanishing from the surrounding countryside.

Cultivating corn was usually women and children's work. The one-horse powered cultivator didn't get all the weeds, and occasionally covered a corn stalk. The second woman using a hoe would uncover and straighten corn plants and remove remaining weeds. The horse wears a fly chaser and ear- and neck-protectors to help keep him safe from pesky insects. *Schwenkfelder Library and Heritage Center, H. Winslow Fegley, photographer*

Planting seeds for grain crops has been done in many ways over the years. Costumed interpreter, Steve Miller, demonstrates hand broadcasting, which has been done since time immemorial. Another old technique is controlled planting by hand. By the mid-nineteenth century there were many implements that could help, such as the "Seed Sowers" available from Benson Maule & Company of Philadelphia, advertised in the *American Agriculturist* in 1883. *The Landis Valley Museum and its Archives; The Mr. and Mrs. Michael B. Emery Collection*

"Hired Girls" often worked in the fields during harvest, as did their mistresses. "Ripened corn was cut, stacked in … shocks and tied so that the wind would not blow them over. This allowed the stalks to dry and the corn … to harden. The shocks were pulled over onto the ground. Huskers knelt … to husk." Once the corn was removed, the remaining stalks were to be used as animal fodder "… tied into sheaves with tar-soaked cord, *Daer-bendel*, which can be seen at the worker's waist." Other husking was done in the barn or farmyard, where a young hired woman, probably a Mennonite, is shown husking at Landis Valley. *Schwenkfelder Library and Heritage Center, H. Winslow Fegley, photographer; Henry K. Landis, photographer, Landis Valley Museum Archives*

A team of five mules provide the power as an Amish farmer harvests corn in contemporary Lancaster County. *Pennsylvania Dutch Convention and Visitors Bureau*

Cornstalks were brought from the field, "… fed into a powered fodder cutter which chopped … [them] into small pieces." An elevator loaded the resulting "chop" into an old-style stone and wood silo where the silage would ferment and be preserved in its own juices. *Schwenkfelder Library and Heritage Center, H. Winslow Fegley, photographer*

Harvesting rye at Landis Valley. Interpreters demonstrate several steps of the arduous process. The grain is cut by hand with sickles or scythes, then raked and loaded onto carts. It would then be taken to the barn for flailing and winnowing. *Landis Valley Museum*

After the wheat, rye or oats harvest, the grain had to be separated from the stalks. In pre-industrial times this was done by first flailing the grain with flailing poles. Then the grain would be winnowed to separate out bits of straw and chaff. This process would change completely with the use of threshing machines and fanning mills. *Landis Valley Museum Archives*

Taking a rest from working on a baling crew, members of the Schott family, Leonard, Joseph, and Leonard, Sr., of Cornwall Township, Lebanon County, pose for the camera inside their (still-standing) barn, circa 1920. The straw, a byproduct of threshing grain, would be machine compressed, formed into bales, and neatly stored until it was needed for animal bedding. *Joe Schott Collection*

Steam-powered threshing machines enabled farmers to mechanically thresh grain brought in from the fields. The shoot blew grain into granaries. Straw waiting to be baled is piled in the background. The whole threshing crew, including the cooks and the farmer, posed for this early twentieth century photograph. *Schwenkfelder Library and Heritage Center, H. Winslow Fegley, photographer*

Northern Berks and western Lehigh Counties were the center of the potato belt, especially Heidelberg, Lynn, and Washington townships, although the crop was grown elsewhere as well. After they were exposed using a potato digger, the tubers had to be collected by hand. Henry K. Landis photographed the harvesting near Landis Valley. *Schwenkfelder Library and Heritage Center, H. Winslow Fegley, photographer; Landis Valley Museum Archives.*

Apple butter, a very popular Pennsylvania German preserve, is made by cooking down cut apples in apple cider along with sugar. Almost always done out-of-doors in a copper boiler, it requires hours of stirring to keep the thick mixture from burning. Henry H.'s diary has several notations of making apple butter on his farm. *Landis Valley Museum*

6th PRIZE—POTATOES
467.54 Bushels

RAYMOND S. RINGER
Allentown (R. D.), Pa.

"I want to say right here that if I was ever tempted to use another Fertilizer I am cured of the notion now. After a crop like this last one in a season that for most of the time my ground was almost desert-dry, I am committed to Animal Base Fertilizer in general — and REICHARD'S in particular. They have certainly made a booster out of yours truly

Raymond S. Ringer

7th PRIZE—POTATOES
465.38 Bushels

ROY RINGER
Allentown, (R.D. 1), Pa.

I used a 1000 lbs. of Reichards' 4-8-8 to the acre and found that the results were more than satisfactory; in fact I have been more than satisfied since I started to use Reichards three years ago.

Roy Ringer

Similar to his Brother Raymond his results are remarkable when considering that the season was very dry. The size of his potatoes was exceptional for the fact that they were planted in 30-inch rows.

The Ringer Brothers of Lehigh County were prizewinners in a 1936 contest sponsored by Robert A. Reichard, Inc., makers of "Reichard's Animal-Base Fertilizers." Champions in agricultural contests often run in families. The first prizewinner was also a Dutchman, Hearl M. Long of Lehighton – but he was not as photogenic. *The Cottage Collection*

Hemp was once a common crop in Dutch Country, but it largely disappeared after the Civil War. This ancient fiber and seed crop, however, continued to be grown elsewhere in the United States until after World War II. *U. S. Department of Agriculture Yearbook, 1901*

Fruit growing has always been important in the region. Apples, pears, and stone fruits (cherries, peaches, plums, and nectarines) continue to be raised. Locals often discuss the merits of York County peaches versus those grown elsewhere. The 1940s postcard touts Franklin County fruits as part of a "Pennsylvania Has Everything" promotion where the "Land of Penn 'Invites You'." The Landis family grew mostly apples, but they also had peach and cherry trees. *The Cottage Collection*

Linton Park (1826-?) painted *Flax Scutching Bee*, c.1860, probably in Clearfield or Indiana County. While the buildings are of German log construction, their form is more British Isles. The process of separating flax fibers from the stalks is shown by the man on the left using a flax break. Others would use paddles to hit handfuls of fiber and stalk against a stump. See the women on the right. At this frolic or bee there seems to be more mischief than work. Perhaps unseen alcoholic beverages were the cause. *The National Gallery of Art, gift of Edgar William and Bernice Chrysler Garbisch*

A mule-drawn tobacco wagon loaded with fresh cut plants stands in front of a tobacco barn. Open slats in the barn allow air to circulate freely to encourage even drying of the leaves. *The Cottage Collection*

Tobacco was Henry H. Landis's favorite crop because it was so profitable. Son Henry K. photographed these hired men as they harvested the crop at Landis Valley early in the twentieth century. *Landis Valley Museum Archives*

Dwight David Eisenhower (1890-1969) was born in Kansas of Pennsylvania German, mostly River Brethern, ancestry. When he retired from the military, he and Mrs. Eisenhower bought the only home they ever owned, a Pennsylvania German farm near the Gettysburg battlefield in Adams County. During Eisenhower's Presidency, the farm was visited by major world leaders and many of the rich and famous to whom he proudly showed his agricultural paradise. It was America's most renowned Pennsylvania German farm; it is now a museum. *The Cottage Collection*

A Harvest Service or a "Harvest Home" was widely celebrated in Lutheran and Reformed churches before the Second World War; later it would slowly languish. This altar, c.1910, was decorated with the fruits of harvest supplied by local farmers. The custom, happily, is now being revived, with most of the produce today often coming from suburban back yards, rather than from farms. *The Cottage Collection*

In summer, a whole range of luscious local fruits and vegetables, including peaches and string beans, are sold at the Tomato Barn's roadside stand in Washington Boro, along with their signature tomatoes. *Lancaster-York Heritage Region*

give and bequeath unto my said beloved wife Elizabeth the following articles and things to be delivered to her yearly and every year as long as she is my widow, by my herein after named son Jacob Landis, his heirs or assigns, in proper time and season in each year out of the lands and premises herein after devised to him and them that is to say; ten bushels of good clean oats, four bushels of good potatoes, and one hind quarter of good fat beef, weighing not less than one hundred pounds including the kidney tallow; as much apple butter and as many summer and winter apples, and other fruit growing on the herein after devised premises as she may need; and as much fuel, coal and wood, as she may want from time to time for her use, the wood to be cut short and split find fit for her use and laid or placed together with the coal near the door of her said dwelling house convenient for her to get it. And my said son Jacob Landis his heirs or assigns shall feed and pasture, stable and litter my said wife's one cow for her use and benefit during the time aforesaid on the premises hereinafter devised to him and them, like his or their own; and shall supply her her hogsty with litter and clean the same, and manure her part of the garden, from time to time during the above mentioned period as often as necessary; and shall furnish her with a safe and good horse creature. ...

The garden was always protected, historically, with a pale or picket fence, which was inexpensive in an age of abundant first-growth lumber. The picket fence was long lasting and, most important of all, allowed for a free flow of air in the garden, very necessary in the humid summers of Eastern and South Central Pennsylvania. When building a fence, posts were placed in the ground at eleven-foot intervals and an eleven-foot panel of the fence, in Dutch *ein gfach*, was attached to them. The principal function of the fence was to keep animals out, particularly rabbits, but also skunks, raccoons, groundhogs, and especially chickens – all notorious crop destroyers. Deer, the bane of modern gardeners, were no problem for the Pennsylvania Germans, who simply slaughtered them for meat whenever they could. The author's ninety-three-year-old mother-in-law, who grew up on a traditional Dutch farm, which included a fruit orchard, reveals that she never saw a wild deer until she was in her twenties. Garden fences, therefore, did not have to be excessively high. The fence, in most cases, was between thirty-six and forty inches high, and the rough cut pales typically extended into the ground or into a sill board that extended into the ground. Additionally, the pales were close together, seldom more than two inches apart, and the top of each one was cut into a point, symmetrically or on a diagonal, which allowed rainwater to drain easily, preventing the wood from rotting, and also made the fence uncomfortable to scale and unpleasant for a marauding cow to graze over. Pales were attached by two nails, one near the top, one near the bottom, to the supporting rails, and the nails were reused when the fence needed to be replaced. There was generally only one opening into a garden that was hung with a gate, *s daerche* or *s daerli* in dialect. The aperture would be made slightly larger than the wheelbarrow used on the farm.

By the nineteenth century it was common to whitewash the picket fence annually, and on a well-kept farm whitewash would also be applied to the bases of grape vines and the trunks of fruit trees, as well as many outbuildings including milk houses and pig pens.

Some early gardens in Germany had irrigation channels; on this side of the Atlantic we have no record of their use in gardens, although irrigated meadows were common (see Chapter IV.) The figure at the left, in this illustration from a German work published in 1598, is grafting a tree. *The Cottage Collection*

An engraving that appears on the title page of a herbarium published in Frankfurt in 1546 shows a gardener attending to a raised bed supported by stakes that are identical to those still used in traditional Pennsylvania Dutch gardens. *The Cottage Collection*

The prime function of the raised beds was to promote easy drainage, but they also allowed for conditioning the soil so that the gardener could grow a larger assortment of crops than would be successful otherwise. Because of the quick drainage of the beds, cool soil vegetables like English (sweet) peas, lettuces, spring onions, and radishes could be planted earlier than they could in the surrounding fields, where wet conditions would rot the precious seed. If the soil on the farm or homestead was heavy clay, or rocky, one or more beds might be filled with sieved soil, enhanced with sand, so that root crops like carrots could grow to their full potential. Great attention was paid to keeping the soil in beds friable and loose. Cultivation was never done with a plow or a cultivator, but always with a hoe, typically a *zink hock*, "a hoe with two heavy tines spaced three inches apart directly opposite a three-inch wide blade." Missed weeds would be hand-pulled. Gardens were never worked when the soil was wet. When planting rows, boards were often placed across the edging planks so that the gardener could lean into the bed without having to rest the weight of her hand on the soil. Many gardeners preferred not to work in the heat of midday or when plants were wet.

What was grown in the kitchen garden varied with time, but it is important to note that a differentiation was made among garden crops. Staples, such as cabbages, potatoes, turnips, and field peas, were usually grown in a larger secondary vegetable patch known as *die lott* or *s schtick*, where the men were more actively involved. Indeed, the kitchen garden was always "...close at hand so the housewife could keep watch over the beds from the window or door" and was "...on the sunny side of the house." It was most commonly either northeast, east, south or south west of the house. Rarely was it located elsewhere. By contrast, *die lott* was often placed beyond the barnyard. Being larger, and level or on a slight slope, it was commonly plowed with the use of horses or mules. It is, of course, the inspiration for many male-tended gardens in Dutch country today.

Mennonite Hanna Rittenhouse Clemens (1880-1977) of Montgomery County recalls that in her childhood,

> ... we had a garden and a patch. And we raised sweet corn and lima bean and all this kept you busy, you know. Now you can go to the store and buy a pack and dump it in the boiler and it's soon fixed. Then you had to plant them and hoe them and pick them and shell them. It's a little bit more work. But they're good.

Generally the kitchen garden provided at least two crops per bed per year. Sometimes, for example, "radishes were broadcast in the onions just prior to the last hoeing and the seeds worked in by hoeing." Besides those already mentioned, favored garden plantings included peppers, eggplant, tomatoes, cucumbers, and melons. Spinach and fancy cabbages, like crinkled-leaf savoy, were also popular, as were specialty corns, such as popping corn.

A raised-bed garden was photographed by Henry K. Landis in Church Town in Lancaster County in the 1930s. The garden is fitted with boards to protect the soil from compaction when the gardener needed to enter the bed. A row of gladiolus (*Gladiolus x hortulanus*) is in the center (left). The Dutch enjoyed the lush flower spikes that lent themselves to growing in rows like vegetables. Many were also raised for market in truck patches. *Landis Valley Museum Archives*

Many foursquare gardens were bordered with additional beds that followed the perimeter fencing. The fences could then be used to support vines or brambles. Cucumber vines might be trained on the fence, as were hopvines, the dried flowers of which were used in making bread and beer, as well as for a pillow filling that was highly regarded as a soporific for insomniacs. Often too, hops might be grown on high poles that would be taken down at harvest time. These border beds might also be used for small fruits like currants or raspberries, or for medicinal or flavoring herbs.

While the layout of the traditional Pennsylvania Dutch garden is derived from the herb gardens of monasteries, the Dutch did not have herb gardens as we understand them. Those are a creation of the colonial revival of the late nineteenth and early twentieth centuries. Traditional German cookery used few flavor enhancers besides salt and pepper (if available). The most commonly grown seasonings were onions, parsley, sage, and dill, which was highly favored for pickling. Saffron, the product of *Crocus savitus*, an autumn-blooming flower that is notoriously laborious to harvest, was also very commonly cultivated, although today it is grown by few gardeners, most of whom are Mennonites and by one last commercial grower, Robert Martin Keen, on land near Landisville that has been in his family since 1725. The other highly favored condiment plant, horseradish, was rarely grown in a garden because of its invasive nature. The same is true of many members of the mint family, which the Germans brewed into teas and gave great medicinal credits to as well. There were exceptions, of course. Mrs. Clemens recalls that her mother's garden

> ... had that wooly balsam tea, the wooly mint, and we had the red stem mint. And we had sage and we had rue in the garden. That rue was so bitter. It's very bitter. Oh, thyme. We had thyme. We made some thyme tea too. That's good with doughnuts. We drank thyme tea with doughnuts.

The rare Pennsylvania Dutch Roman Catholic sometimes placed a potted rosemary in the center of a foursquare garden, but its significance was religious, associated with the Virgin Mary, not necessarily culinary. Along these lines, Pennsylvania Dutch Protestants often planted an Adam and Eve plant (*Yucca filamentosa*) in the middle of their gardens. However, how this American native plant came to be considered an expression of Protestant faith is not known.

Traditionally, as noted earlier, the Pennsylvania Dutch garden was placed on level or gently sloping ground, although the archaeologically correct, recreated garden at Burnside Plantation, a Moravian site in Bethlehem, Pennsylvania, is on a steep terraced site. Most important was the availability of water, for raised gardens drained so well that they needed to be watered more frequently than did flat gardens. Pennsylvania German farms were invariably laid out on an east-west axis, with the house and barn facing south (as noted earlier) and the garden was usually placed near the kitchen door, where it could share a water source, generally a well or spring, with the house. A garden needs at least one inch of water per week, so the chore could be arduous indeed in times of drought, when it might be necessary to fill and haul buckets from more distant water sources. Watering cans were in use early but considered a luxury until inexpensive machine-made galvanized-metal ones became widely available about 1900.

"Farm House and Yard of the Luth[eran] Orphans' Home" in Topton shows several interesting garden features, including a board sidewalk, tropical plants in large containers placed on the lawn, and a raised bed against the fence. One of the stakes supporting the boarding is clearly visible. *The Cottage Collection*

Other cultures also used raised-bed gardens, but maintenance was often more casual than among the Pennsylvania Germans. The recreated garden at Plymouth Plantation in Massachusetts, an English variant, is surrounded by a high, solid fence, which would stifle air movement if exposed to sultry central Pennsylvania summers. *Author's photograph*

Ellen Gehret [a leading textiles scholar, Alan G. Keyser relates] uncovered a rather unusual method of watering plants growing on hills. An earthen crock was placed next to each hill and a woolen rag was put into the crock with one end in the water and the other end was laid on the hill where the plants were growing. The water then wicked from the crock to the plants. A woolen rag had to be used because no other fabric would wick as well. This was not a common practice, but it was used in the upper end of Montgomery County and Ellen Gehret found two informants who had seen this used when they were younger. North of Boyertown in Berks County there was until several years ago a garden with pale fence around it, and on the fence were hung upside down on the pales many glass jars and earthen crocks. These we always felt were just the applebutter crocks in storage, but they may have been the irrigation crocks and jars for watering by the wick method.

The Pennsylvania Dutch were among the first farmers to barn, pen, and pasture their animals, practices that allowed them to collect manure, which they applied liberally to both gardens and fields. They did not use compost, but they often placed their garden near the pigpen, chicken yard, or rabbit hutches so that these animals could convert less palatable garden produce into manure. Among the most dedicated of gardeners, different manures were, and are, preferred for particular crops. Pig manure was especially favored, but some believed that it made the soil too hard. The problem could be dealt with by adding wood ash from the stove to the soil as well. The best day for this was, of course, Ash Wednesday.

Flowers, per se, had little place in Pennsylvania Dutch kitchen gardens originally, although some plants we consider ornamental were grown for their food value. Nasturtiums, for instance, were often raised for their buds, which were pickled to provide a caper substitute; the tender leaves of calendulas, or "Pot Marigolds" (*Calendula officinalis*) were eaten as a vegetable and the flowers were used as a dyestuff. Occasionally, French marigolds (*Tagetas patula*), which the Germans called, in translation, "stinky flowers," were grown among beans to repel beetles. In this vein, it is important to remember that although the Pennsylvania Dutch, until the twentieth century, had to contend with fewer garden pests than we do today, the problem has always existed. On occasion they used lethal, but organic, substances such as tobacco dust and highly toxic arsenic-based compounds to deter insects, but a good deal of pest control was purely mechanical; the women and children hand picked insects, snails, and slugs, which is still the most effective organic method of pest control.

The traditional Pennsylvania Dutch garden was eventually abandoned because of increasing interaction with the "English" world and the rising importance of commercial seed houses and commercial or truck farming. The final blow was the widespread availability of the rototiller after World War II. When power machinery entered the picture, vegetable gardening among nonsectarian Pennsylvania Dutch appears to have shifted to men.

The Pennsylvania Dutch kitchen garden form was in flux as it declined. A first step was to remove the side paths and to make the garden into a bilaterally symmetric expression. Additionally, gardens would admit more flowers and even shrubs, yielding more food production to *die lott*. Certainly in town and on prosperous farms

The drawing of a six-bed raised garden was done by Susanna Heebner (1750-1818). The house in the background is based on the still-standing Peter Wentz Homestead, now a museum in Montgomery County. Probably taking artistic license, Susanna Heebner portrays the beds filled with flowers, rather than vegetables. She also shows the garden with three gates instead of the customary one. Note the wonderful decoration on the corner posts and the multicolor of the gates. *Schwenkfelder Library and Heritage Center*

more and more vegetables and herbs might be purchased and/or bartered. Henry H. Landis' diary is replete with mentions of garden produce, including parsnips that were gifts from neighbors as well as various purchases including strawberries. Flowerbeds also often existed outside of the kitchen garden itself, especially by the late nineteenth and pre-World War II twentieth centuries. Among Mennonites, especially where the custom of Sunday after-church visits remained strong, flowerbeds were often planted between the house and the garden, so that even the Plainest of the Mennonites could show off their flowers when showing their relatives and neighbors the kitchen gardens.

When herbs for medicinal or culinary uses were grown in the historic garden they were usually planted in one of the perimeter or outside beds, often one nearest to the gate. Similarly, flowers, if present, would usually be grown in the outside beds as well – often on the opposite side of the gate from the herbs.

The following is a list of plants commonly grown in Pennsylvania Dutch gardens, which is adapted from one compiled by David E. Lick and the Reverend Thomas R. Brendle in 1923 when many gardens and informants not available to later researchers were still about. Alan Keyser and others have also added to the list. Of course it is also important to remember that not all of these plants would have been found in any one garden or at any one date.

OUTSIDE BEDS:	
Agrimony	*Agrimonia*, genus.
Asparagus	*Asparagus officinalis*
Asphodel	*Asphodelus luteus.*
Avens	*Geum*, genus; *Geum urbanum*
Black Currant	*Ribes nigrum*
Black Raspberry	*Rubus occidentalis*
Blessed Thistle	*Cnicus Benedictus.*
Blue Vervain	*Verbena Officinalis.*
Boxwood	*Buxus sempervirens*
Calabash	*Lagenaria vulgaris*
Castor Bean	*Ricinus communis*
Catnip	*Nepeta Cataria*
Chamomile	*Anthemis nobilis*
Chives	*Allium Schoenoprasium*
Coltsfoot	*Tussilago Farfara.*
Comfrey	*Symphytum officinale*
Downy Mint	*Mentha alopercuriodes.*
Elecampane	*Inuls Helenium*
Feverfew	*Chrysanthemum Parthenium*
Fumitory	*Fumaria officinalis.*
Garden Balm	*Melisa officinalis*
Gooseberry	*Ribes reclinatum*
Grape	*Vitis* genus.
Ground Almonds	*Cyperus esculentus.*
Hellebore	*Helleborus viridis.*
High Blackberries	*Rubus allegheniensis*
Hop	*Humulus Lupulus*
Horehound	*Marrubium vulgare.*
Horse Radish	*Radicula Armoracia*
Hyacinth Bean	*Dolichos Lablab*
Hyssop	*Hyssopus officinalis.*
Jerusalem Artichoke	*Helianthus tuberosus.*
Juneberry	*Amelanchier*, genus.
Lady Thistle	*Silybum marianum*
Lavender	*Lavendula vera*
Lilac Bush	*Syringa vulgaris*
Lovage	*Levisticum officinale*
Lungwort	*Pulmonara officinalis.*
Masterwort	*Imperatoria Ostruthium*
Meadow Sweet	*Filipendula Ulmarid*
Motherwort	*Leonurus Cardiaca*
Mugwort	*Artemisia vulgaris*
Oswego Tea	*Monarda didyma*
Parsley	*Carum Petroselinum*
Peppermint	*Mentha Piperita.*
Pimpernell	*Pimpinella Saxifraga*
Red Currant	*Ribes vulgare*
Red Raspberry	*Rubus idaeus*
Rhubarb	*Rheum Rhaponticum*
Rue	*Ruta graveolens*
Saffron	*Crocus sativus*
Sage	*Salvia officinalis*
Self-heal	*Prunella vulgaris*
Southernwood	*Artemisia, Abrotanum.*
Spearmint	*Mentha spicata.*
Speedwell	*Veronica officinalis.*
Spurge	*Euphorbia Lathyrus*
Strawberry	*Fragaria*, genus.
Sweet Cicely	*Osmorhiza*, genus.
Sweet Marjoram	*Origanum Marjorana*
Tansy	*Tanacetum vulgare*
Thyme	*Thymus serpyllum*
Tithonia	*Tithonia rotundiflora*
Valerian	*Valeriana officinalis*
Wormwood	*Artemisia Absinthium*

FOUR MAIN BEDS:	
Beans	Bush varieties and occasionally pole beans
Borage	*Borage officinalis*
Cabbage	*Brassica oleracea*, *Capitata* Group
Caraway	*Carum Carvi*
Carrot	*Daucus Carota*
Catchfly	*Silene inflata*
Cauliflower	*Brassica oleracea*, *Botrytis* Group
Celery	*Apium graveolens*
Coriander	*Coriander sativium*
Corn Salad	*Valerianella*, genus
Cucumber	*Cucumis sativus*
Endive	*Cichorium Endivia*
English Peas	*Pisum sativum*, and varieties
Fennel	*Foeniculum vulgare*
Ground Cherry	*Physalis*, genus
Job's Tears	*Coix Lachryma-Jobii*
Kohlrabi	*Brassica oleracea*, *Gongylodes* Group
Lentils	*Lens esculenta*
Lettuce	*Lactuca sativa*
Mammoth Pumpkin	*Cucurbita Maxima* and var.
Multiplier	*Allium Cepa* var. *multiplicans.*
Onion	*Allium Cepa*
Orach	*Atriplex hortensis*
Parsnips	*Pastinaca sativa*
Peppers	*Capsicum annum* var.
Popcorn	Var. of *Zea Mays*
Potato	*Solanum tuberosum*
Radish	*Raphanus sativus*
Red Beed	*Beta vulgaris*
Savory	*Satureja Hortensis*
Spinach	*Spinacia oleracea*
Tobacco	*Nicotiana*, genus
Tomato	*Lycopersicum esculentum*
Top Onion	*Allium proliferum*
Turnip	*Brassica Rapa*

ORNAMENTALS AND FLOWERS:

Amaranth	*Amaranthus* tricolor
Amaranth	*Amaranthus*, genus.
Aster	*Asternovai-angliae*
August Lily	*Hosta plantaginea*
Bergamot Mint	*Mentha peperita var. citrata*
Birthwort	*Aristolochia Clematitis*
Bleeding Heart	*Dicentra spectabilis*
Calendula	*Calendula officinalis*
China Asters	*Callistephus chinensus*
Chrysanthemum	*Chrysanthemum*, genus
Columbine	*Aquilegia*, genus
Corn Poppy	*Papaver Rhoeas*
Creeping Boys and Girls	*Sedum acre*
Daffodil	*Narcissus pseudo narcissus*
Dahlia	*Dahlia variabilis*
Day Lily	*Hemerocallis fulva*
Dusty Miller	*Centaurea* and *Cinervaria*
French Marigolds	*Tagetas patula*
Garden Fuchsia	*Fuchsia*, genus
Garden Orpine	*Sedum purpureum*
Garden Pinks	*Dianthus Caryphyllus var.*
Geranium	*Pelargonium*, genus.
Grape Hyacinth	*Muscari botryoides*
Hollyhock	*Althea rosea*
Honesty	*Lunaria biennis*
Houseleek	*Sempervivum tectorum*
Hyacinth	*Hyacinthus orientalis*
Immortelle	*Helichrysum*, genus.
Iris	*Iris*, genus.
Larkspur	*Delphinium*, genus.
Leopard Plant	*Farfugium grande*
Lily-of-the-Valley	*Convallaria Majalis*
Madonna Lily	*Lilium candidum*
Moneywort	*Lysimachia Nummularia*
Nasturtium	*Tropaelua Maias*
Peony	*Paeonia officinalis* and varieties
Periwinkle	*Vinca minor*
Petunia	*Petunia*, genus.
Phlox	*Phlox Drummondii*
Poppy	*Papaver*, genus
Prince's Feather	*Polygonum Orientale*
Red-Hot-Poker-Plant	*Kniphofia aloides*
Rose Moss	*Portulaca grandiflora*
Rose-of-Heaven	*Agrostemma Coeli-rosa*
Snapdragun	*Antirrhinum majus*
Stocks	*Matthiola*, genus.
Summer Cypress	*Kochia scoparia*
Sweet Alyssum	*Alyssum maritimum*
Sweet William	*Dianthus barbatus*
Tiger Lily	*Lilium tigrinum*
Violet	*Viola*, blue species
Woodland Tulip	*Tulipa sylvestris*

CENTER BED

Adam & Eve Plant	*Yucca filamentosa*
Crown of Thorns	*Berberis* – a cultivated dwarf form
Hollyhock	*Atheo rosea*
Pansy	*Viola* tricolor
Rosemary	*Rosemarius officinalis*
Saffron	*Crocus sativus*

The Pennsylvania Germans have been traditionally both religious and superstitious. These were the people, please remember, who gave us the Easter Bunny. Accordingly their beliefs and traditions often governed the dates and ways in which they planted. These mystic rules were followed both by women in their kitchen gardens and men in *die lott*. It is fair to assert that traditionally it is probable that every housewife and her husband used some sort of system based on the church calendar and/or the signs of the zodiac in their gardening. Even today many people who prepare their normative looking gardens with a rototiller use such a system. Various old farmers' almanacs continue to be published for and purchased by the traditionally minded.

The "Farmstead of Jacob S. Hunsecker" was drawn c.1869-1875 by Charles H. Wolf (?-?). A view in southeastern Pennsylvania, the barn complex demonstrates that this was a very prosperous farm. Note the ramp leading into the bank barn. The large house and its ancillary structures are clearly drawn. The yard surrounding the house is fenced as is the precisely drawn garden. *Watercolor and ink on paper, 23 7/8 x 31 3/4 inches. Collection American Folk Art Museum, New York; Promised gift of Ralph Esmerian, P1.2001.61*

Before getting to the specifics of the systems, keep in mind that it was absolutely forbidden to work in the garden on the Sabbath, nor on Ascension Day. It was often debated as to whether or not it was permissible to work on Good Friday and some of the pious would not work on "... Saturday or at least on Saturday afternoon, because it was too close to Sunday." Gardeners also gave themselves every chance to imbue praise of their cleanliness and order with religion. Garden tools were, of course, to be kept clean and shining and shovels were often oiled to prevent them from rusting. Fastnacht grease [used up on Fastnacht day or Shrove Tuesday] was believed not only to preserve the shovel better but to also make it slide easily into the soil and to kill harmful insects in the garden.

Many first person accounts of growing plants following a mystical system were gathered by folklorist and scholar Alan G. Keyser. His reports are clear, direct, and often intriguing.

> From Sam Heller we were able to collect his mother's system. This method of planting by the moon seems to be the most commonly followed. Her two signs for planting in the garden were *zwilling un wog.* [Gemini and Libra] Sam said that his mother told him that "*zwillingdut alles dobble,*" [The twins double everything] and in "*wog grickscht alles wider zurick.*" [In the scales everything is returned.] From another source we have the type of seeds to be planted on the above days. Things which yield above the ground such as beans and pickles should be planted in the twins because it is the ascent of the moon. Things which grow below the ground such as red beets, radishes, and potatoes should be planted in the scales because it is in the descent of the moon. Sam's mother also liked to transplant her flowers on *der blumme daak.* [The flower day, i.e. the sign of the Virgin.] This sign was considered to be good for flowers only, and any vegetables planted in the virgin would only bloom. Nothing was ever to be planted in *der schitz,* [Sagittarius—the bowman. There is one exception to this common belief – pole beans should be planted on the bowman so they will go up the pole] because it would "shoot" to seed. Nothing was planted *uff fisch odder wasserman* [on Pisces or Aquarius] because it would be too wet. Sallie Snyder told me that if you plant red beets in the sign of the fish they will get stringy, and the seeds planted in waterman will rot. Do not plant anything in Scorpio or Capricorn. (Here I have not been able to find reasons or explanations for these beliefs.) *Grebs* (the crab) is also not a good sign for planting, and Emma Becker of Washington Township, Berks County, gave the reason. The "crab walks backwards so the plants will grow backwards or insects will eat the plants." No onions or radishes should be planted in *Leeb* (Leo) because they will have a strong flavor. You may plant in *schofbock* but there is no guarantee of favorable results. Sam Heller did not make any comment on the sign of Taurus nor was I able to find anyone else who could.

The Gruber Farm in North Heidelberg Township, Berks County, was photographed by C. G. Blatt in 1902. Visible is a slightly dilapidated kitchen garden, but there is a big fine haystack in the barnyard. Farmer Michael A. Gruber is pictured "... the field in which he is standing is known as the lower hill field, near the top of which the view was taken."
Private Collection

Charlie Miller of Bechtelsville uses a different system of planting by the signs of the moon. His appears to take advantage of some of the weak points in the previous system, and also utilizes its strong points. He, for instance, plants his lettuce and radishes in the "waterman" and "fish" so they will become nice and crisp. Charlie also uses the last day of the "blooming lady: (Virgo) for some of his crops which should bloom in order to bear. This includes sugar peas and cucumbers, but he adds that if you plant on the first *jungfrau* (Virgo) you will get nothing but flowers. He also stated that he likes to plant his potatoes in either the "stonebuck" (Capricornus) or the "scale," and was quick to add that the potatoes planted in "stonebuck gave more than in the scales."

Lizzie Heller of near Walnuttown, Berks County, was able to add this to the list of favorable signs for planting. "*Blans in wider no grichscht wider.*" [Plant in Aries and you will receive again.] She also prefers to plant in *schteebock*, [Capricornus] and her beans in the "twins." As everyone else Mrs. Heller declared that the bowman was a poor sign for planting because "*es schiesst in die heh.*" ["It shoots up."]

Keyser points up too that many gardeners – yesterday and today—ignored many subtleties and employed "...a very simple system, and use nothing more than the ascent of the moon for planting crops which yield above ground, and crops yielding below the ground in the descent."

A multi-generational farm along the Maiden Creek in Berks County had at least two kitchen gardens. Grapes grow lushly on an arbor leading to the house at mid-left. Three children pose on the porch in a hand-tinted photograph, c.1890. *The Mr. and Mrs. Michael B. Emery Collection*

Even the most casual of the sign-based gardeners were aware of St. Gertraut's (or St. Gertrude's) Day, which was celebrated on March 17. Traditionally this was the day you planted onions and especially cabbage. "If the cabbage seed were sown on any other day it would have become bug infested." After the great Irish immigration of the 1840s, almanac writers, even German language ones, started using St. Patrick as the marker for March 17. And while the date of March 17 retained its sacred quality for German gardeners, its connection to their past became more obscure.

In the past two decades, there has been a revival of appreciation for the raised-bed garden, both by organic gardeners and by historic sites and museums. A panel of the popular comic strip "Hi and Lois" (Sunday, October 23, 2005) shows Lois working at a raised bed garden. How much more popular culture can you get? Organic gardeners appreciate the virtues of a garden that allows for the use of manure or compost rather than chemical fertilizer and a form that provides substantial yields in a small area.

The William Cleaver Farm in Oley Township, Berks County. The lush meadow in the foreground is in a flood plain. The very large garden probably fed more than one family. The large barn is decorated with hex signs. The farm was infamous as the site of the Susanna Cox murder. "Susanna Cox was a maid on this farm and lived in the small stone house, according to tradition. The story of her act of infanticide and subsequent execution by hanging in 1809 in Reading was celebrated on one of the most frequently reprinted song broadsides among the Pennsylvania Dutch." *Schwenkfelder Library and Heritage Center, H. Winslow Fegley, photographer.*

A traditional farmhouse and garden in Salisbury Township, Lancaster. Tom Martin, a Landis Valley Museum interpreter, worked in the 55' × 55' garden until recent times. The photograph, however, is early twentieth century. *Tom Martin*

Articles on raised-bed, usually boarded, gardens are common in many publications – especially those specializing in organic gardening. For modern home gardeners who want to create a raised-bed garden there are several caveats. First and most important is that you never use pressure treated lumber for enclosing your beds. Treated lumber often contains arsenic or other toxic substances that can easily contaminate your soil. Modern pine or oak boards tend to rot quickly. Cedar is a long-lasting wood that can be used as can redwood, but here the problem is encouraging the cutting of these irreplaceable giant trees. Similar questions can be raised about the use of some tropical woods, including fine teak. Very durable, but not as aesthetically pleasing as wood, are plastic boards made for the purpose from recycled milk jugs. They are generally available by mail order from advertisements in organic gardening magazines.

For those who are mechanically involved, and do not want to dig their garden each season, the chore can be handled with the new lightweight mini tillers. While museums should be scrupulous about dimensions and path surfaces, home gardeners can be a bit more relaxed. You might want to use an eight-foot multiplier instead of an eleven foot one. Sections of prefabricated picket fences come in eight-foot sections. As a caution, if you use prefabricated fencing you will also want to line it with chicken wire on the inside. Modern picket fences allow too many critters in. Also, you might want to make your paths wider for easier maneuvering by you and your garden cart. Additionally, most modern gardeners will want to use mulch on the paths. Today you have the extra advantage of using black plastic or horticultural cloth under your tanbark or other appropriate mulch.

Landis Valley Museum has been a pioneer in the revival of the traditional Pennsylvania German garden, and it is also the home to the Heirloom Seed Project, which makes traditional Pennsylvania Dutch varieties of vegetables and flower seeds widely available. The image of actual historic gardens at Landis Valley is blurred. Gardens by their very nature are transient. Those with little in the way of substantial structure leave few traces, especially in land that has been disturbed over many years.

As always, early Henry H's diary is of great value. The diary has many references about digging the garden or "Emma's garden." Henry was occasionally the digger, but the work was more commonly done by a hired man. "Billy is digging garden" [April 10, 1889] Henry wrote on a fine warm Saturday day in April 1881, "...I dug garden I finished two large beds the boys hauled in the manure". There are also frequent references to whitewashing the fence. "Geo whitewashed the garden fence on the inside" [April 17, 1881]. Pales for garden fencing were also mentioned on a regular basis. For example, Henry and "the boys" went to Lancaster on April 27, 1876, where among his other purchases were "a Keg of nails and some 300 four foot pails [*sic*] from B. B. Martin." Elsewhere he values pales at two cents each.

The garden of the miller's house at Pleasant Valley in Bern Township, Berks County, was extensive. It also served as a drying yard. Note the wash line over the walk. Retaining boards surround the beds. Onions and lettuce are easily identified, as is the rhubarb along the fence. A large rose bush is near the gate and a grape arbor to the right. *Schwenkfelder Library and Heritage Center, H. Winslow Fegley, photographer*

The diary has very few references to what Emma actually grew in her garden. Henry bought "a pack of Cabbage seed for Emma" on April 14, 1914—no doubt she did not believe in the magic of St. Gertrude's Day. She waited a whole month after the appointed day to plant her cabbage. And on September 1889, "Emma took out her sweet potatoes." Henry often mentions picking blackberries and raspberries for Emma. Were these berries within the confine of her garden, or outside? We simply do not know. There were also grape vines near the house, but we do not always know their precise location.

Keeping the yard around the house in order was one of Henry's responsibilities. He notes his cutting the lawn and raking the grass clippings, which Emma claimed for her "young chicks." More often it was a hired man who did yard work. "Hen was cleaning up the yard and I worked," Henry recorded July 19, 1876.

Existing photographs show that there were freestanding flower beds, independent of Emma's garden. In May 1876 Henry notes that he "...Made a flower Bed [*sic*] for Emma." The most poignant references to independent gardens relate to Henry's decorating Nettie May's grave when he tearfully laments the decoration gathered from flowers Nettie herself had planted.

The house lot of photographer C. G. Blatt, c.1885, in Bernville, Berks County, was intensely farmed. In addition to the vegetable garden at left, there are also rabbit hutches and dove cotes, *Taubenschlacken*, as well as a young orchard, a barn, and a smokehouse. *Private Collection*

There was almost certainly a large vegetable patch, *die lott*, cultivated by Henry and his crew. Like the kitchen garden, this one was probably fenced. It was here Henry no doubt grew his potatoes. On several occasions he notes harvesting his sweet potatoes – which were clearly distinct from Emma's. May 30, 1887, Henry notes in his diary that, "Hen and I planted pumpkin seed up to 9 o clock". We don't know if Henry was superstitious but it was often believed that, "...unruly pumpkin and squash spirits cause chaos in tidy German Gardens." In July of 1883 Henry records "I gave H. L. Brackbill some Turnip Seed and sowed some myself". This suggests that Henry had saved these seeds from the year before.

The most intriguing of the entries in Henry's diary about a garden practice appears on January 22, 1876. Henry notes one of his father's demented acts — this one at the Landis Valley Hotel. "I went to the tavern Henry L. Landis was there Pap left a big pile right in front of the proch [*sic*] I took the shovel over and carried it to the garden" Was this just how one dealt with an unfortunate accident – or did the Dutch also commonly use what the Chinese referred to as "night soil" as fertilizer? Perhaps we shall never know.

In addition to the traditional garden, there have also been some large scale ornamental gardens developed by the Pennsylvania Germans. The Moravian contribution has already been discussed in Chapter II. Another group with a rich garden-building heritage was the Harmonists, who had broken from the Lutheran Church and had been badly persecuted in Europe.

George Rapp (1757-1847), the son of a small farmer and vinedresser, gathered together a group of 300 Pietist families who shared his beliefs. In 1803 Rapp, at age forty-six, sailed to Baltimore, Maryland, and subsequently bought 5,000 acres of virgin land just north of Pittsburgh, Pennsylvania. On July 4, 1804, about 300 of his followers arrived in Baltimore, followed by another 300 who landed in Philadelphia. In 1805, they formally united as the Harmony Society, and in their first year they built about fifty log houses, a gristmill, and a barn and workshops, and they started clearing land.

Two women stand in the garden gate, *s daerli*, in front of the Mohn House, which was built in the eighteenth century. *Schwenkfelder Library and Heritage Center, H. Winslow Fegley, photographer*

Harmony was soon to be prosperous, but it was inland from a navigable river and it was poor wine growing country. As a remedy, in 1814, the community bought 30,000 acres in Indiana's Wabash Valley, selling their original colony for $100,000. They also called their new settlement "Harmony," which also prospered as more settlers joined them. In Indiana, however, malaria was rampant and their neighbors unpleasant. Accordingly, they sold the community to Robert Owen (1771-1858), the famed English reformer who established his famed New Harmony colony there.

Now the community bought land at Economy, north of Pittsburgh, Pennsylvania, near their first settlement. By 1825, the entire community would be settled in their new town on the Ohio River where they would grow rich. A comfort-minded people, they were famous for their winery.

Each family house had its own vegetable garden and adjoining the Great House, Rapp's House, there were elaborate ornamental gardens, rife with mystical symbolism and celebrated for their beauty. Philosophically this was perhaps closer to the Landis brothers aesthetic.

While George and Henry collected some garden related tools and implements, they never consciously recognized the garden itself as an artifact to be preserved. In their lifetime it was just, perhaps, too commonplace.

The only known document about the gardens at Landis Valley as a museum in the brothers' day is:

"List of Herbs
Landis Valley Museum Herb Garden
Planted by the
MENS GARDEN CLUB
Of Lancaster
Compiled by Mrs. Walter F. Kaufman"
And it bears a stamp:
"Landis Valley Museum
H. K. Landis
Geo. D. Landis
LANCASTER
PA"

Pipe smoking Mrs. Jacob Sauder stands in her front yard, whose gate and fence are "… primitive random – with riven pickets." Part of a small raised bed is seen to the right of the door. A rose of Sharon (*Hibiscus syriacus*) blooms behind the elderly woman. *Schwenkfelder Library and Heritage Center, H. Winslow Fegley, photographer*

Some German gardens were built on terraced, rather than flat, ground. Portions of this garden appear to be on more than one level. An apple orchard is at the right. *Mr. And Mrs. Michael B. Emery Collection*

The garden planted under their supervision near the reconstructed tavern building was a colonial revival herb garden with thirty-seven varieties of plants ranging alphabetically from Angelica to Woodruff, and including elecampane and gas plant, along with the more expected mints, tansy, and sage.

Each entry is keyed "a annual b biennial p perennial: attractive to bees CU Chief use in medicine and cooking today according to the findings of the Brooklyn Botanic Garden. m medicine c culinary". Among the very un-Dutch information we are given is that, "Roman Wormwood" is used in making "Vermouth" and that chamomile is good for flavoring sherry!

The most unusual inclusion on the list is:

> Helleborus niger. Black Hellebore. Christmas Rose. Rare even in the old herb gardens. It was classed with the strewing herbs. In the garden it was usually planted near the house because it was believed that no evil spirit would enter a dwelling near which these plants were grown. It is a perennial of very long life. It is not uncommon to have a clump over thirty years old.

Always difficult to grow in this region there is no Christmas Rose on the property today.

The most elaborate entry was devoted to Rosemary:

Rosemary. Rosmarinus officinalis. P * Not hardy here. Wood shrub, beautiful narrow green leaves. Tiny bright blue flowers. Spicy nutmeg odor. Warm balsam taste. According to old writers this plant is full of virtues—its leaves placed under the bed prevent the sleeper from having evil dreams; its flowers placed among clothes and books prevent moths from destroying; if the leaves are boiled in white wine and the face and brows washed therewith "thou shalt have a faire face"; and the odor of a box made of the wood of rosemary was said to preserve youth. In mediaeval times Rosemary was used as a garland on the most important dish at festive occasions [*sic*]. The leaves are used by the Italians today for flavoring. CU. A strewing herb, garnish, and for flavoring.

Note the information that the herb is used "... by the Italians today for flavoring." It certainly is not a staple of Dutch cookery. Rosemary today is grown in the Heirloom Seed Project's display garden.

George and Henry's spiritual heirs, the Landis Valley Associates and the Curators of Landis Valley Museum, have remedied the Landis gardening lapse and the grounds are dotted with appropriate gardens. We would hope that the Landises would find at least some of them familiar.

Grapevines grow along the porch of a double-door German farmhouse in Lancaster County. Note that, historically, the Dutch did not use foundation plantings. A woodbox is on the porch (left). Less affluent farmers, as in this case, would often apply fresh paint to only the exterior of the first floor. *Henry K. Landis, photographer, Landis Valley Museum Archives*

A woman works at hoeing an unfenced truck patch or *die lott* on a poor farm. The small Germanic stone end-chimney house form was familiar in Berks County. Note the grape arbor attached to the house. It provided fruit as well as a pleasant shady area where the housewife could work on a sunny day. A smokehouse is to the right. This impressionist painting was done by Ralph D. Dunkelberger (1894-1965). *Schwarz Gallery, Philadelphia*

Cabbage grows in a truck patch or *die lott* near a farmhouse complex in an unsigned oil painting, c. 1890. *The Cottage Collection*

The German garden followed the culture wherever it went. A proud family poses in their fenced garden near Lamartine, in the Western Pennsylvania oil fields, c.1900. The bottom of the fence is protected by a sill board, which would keep burrowing animals out. The man in the middle stands next to a rhubarb (*Rheum Rhabarbarus*) patch. Raised garden beds are left of him. A peach tree grows nearby. Athletic bystanders also got into the picture. *Pennsylvania State Archives, Pennsylvania Historical and Museum Commission*

The Berks County Alms House fed its inmates with food raised on their own farm and gardens. The greenhouses and neatly maintained garden beds are seen in the oval. This view, on metal, was painted by Charles C. Hoffman (1820-1882) in 1878. *National Gallery of Art, Gift of Edgar William and Bernice Chrysler Garbish*

A modern schematic illustration of the basic four-square Pennsylvania German garden clearly illustrates the form's composition. Note that the central bed, which typically grew ornamentals, was not usually raised. *Lancaster-York Heritage Region*

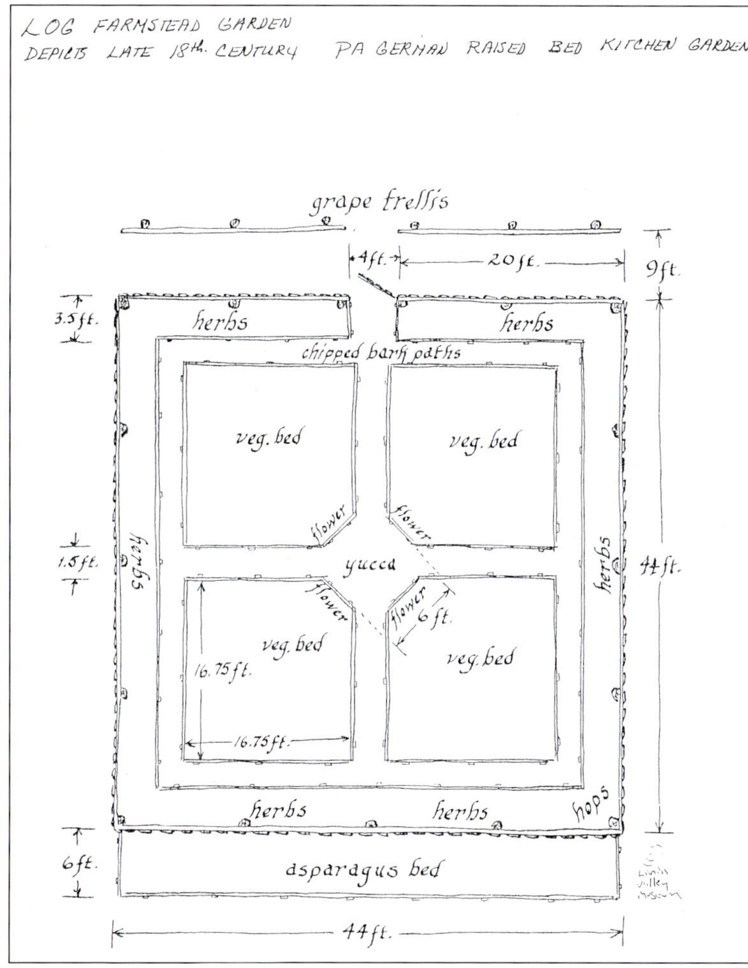

A garden plan for the newly restored four-square garden at the Log Farm complex. The chipped bark on the path is for modern convenience. The perimeter beds are planted with too many herbs to be strictly accurate, but they reflect visitor interest. All the plant varieties used, however, are accurate to the period. This diagram is easily adapted for use by the home gardener. *Landis Valley Associates*

A modern raised-bed (without boards) four-square garden in the Palatinate is very much oriented towards flowers. A lilac, trained as a small tree, is coming into bloom. *Nadine Steinmetz, photographer*

The now abandoned garden at Donegal Mills Plantation near Mount Joy, Lancaster County, featured an especially elaborate center bed. The structure to the left is a tool shed. *Author's photograph*

Steve Miller, the Director of the Landis Valley Museum, lives in an expanded nineteenth century German farmhouse and he carefully tends a four-square garden filled with vegetables and some flowers. Peas are drying off in the back right and tomato plants thrive at the left. Steve's center bed is bordered in ornamental concrete blocks, which replace the tractor tire he used earlier – a common garden ornament since the 1920s. Located in Conestoga in southern Lancaster County, Steve, like the Landis brothers, occasionally finds arrowheads while working his garden. *Steven A. Miller, photographer*

What not to do when planning a German garden is shown in this modern Lancaster County garden – now redesigned. The paths covered in unsympathetic crushed stone surround too small raised beds. *Author's photograph*

Almanacs were of great importance to the Germans. They provided needed advice on planting times and festivals. Their covers and frontispieces were akin to trademarks. (Samuel?) Leech engraved the Almanac cover, with its farming vignette, for John Tobler's English language (but German oriented) *The Pennsylvania Town and Country Man's Almanac*, published in Germantown by Christian Sower (Saur) II in 1756. (See Chapter II). *Winterthur Library, Printed Book and Periodical Collection*

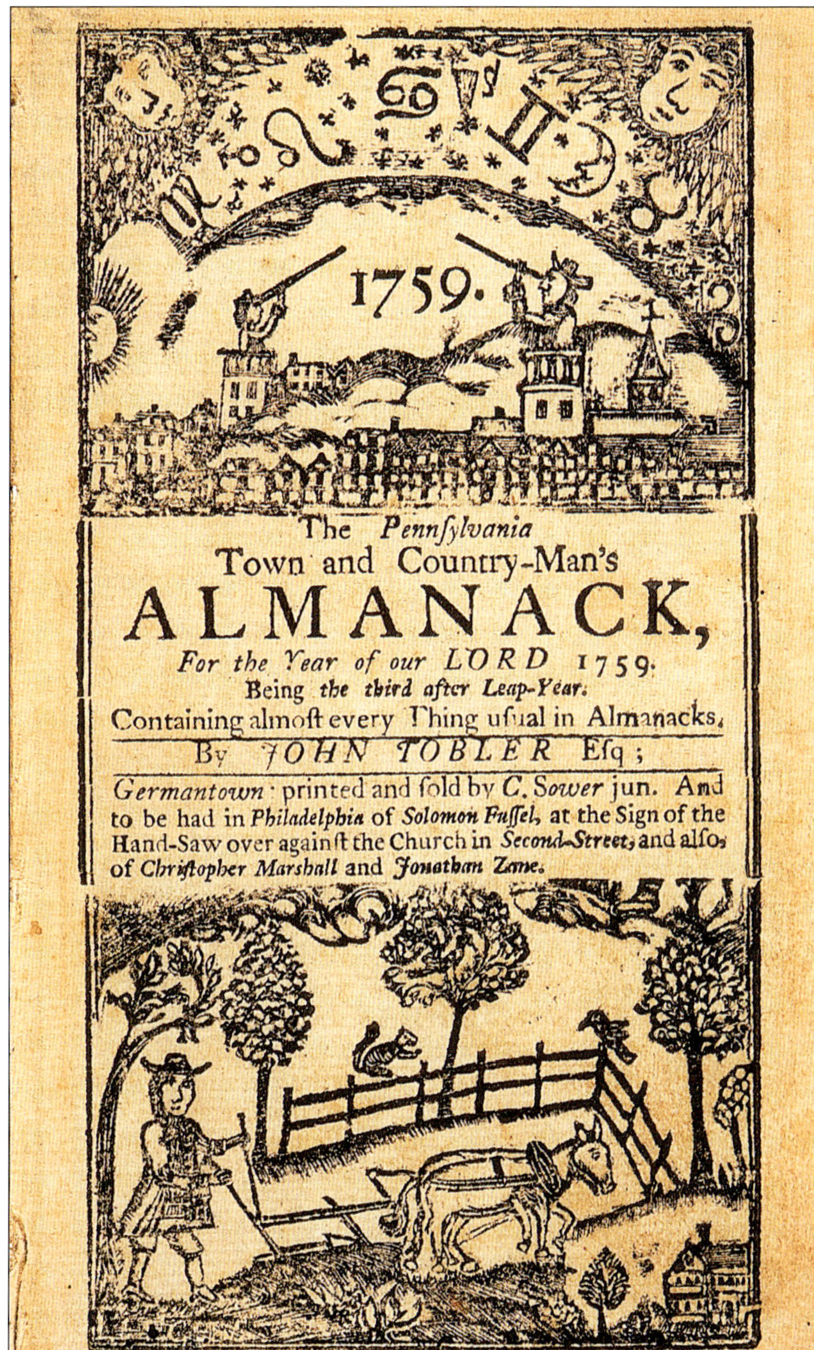

The Pennsylvania Germans have given American culture the Easter Rabbit and the decorated eggs – both of which are embedded in middle-European lore. The earliest known image of the Easter Rabbit in America is attributed to Conrad Gilbert (1734-1812) of Berks County. Images of baskets of eggs are extremely common as are highly imaginative Easter cards imported from Germany. The Pennsylvania Germans also introduced the Easter egg tree. Most elaborate was the "Easter Tree Putz," which belonged to the Gaumer family of Allentown. It contained over 1,000 eggs and related items. Contemporary front yard displays replete with plastic eggs and bunnies are a common Dutch Country sight. *Abby Aldrich Rockefeller Folk Art Museum; Mr. and Mrs. Michael B. Emery Collection; The Cottage Collection; Author's photograph*

Peonies (*Paeonia peregrine*), pictured on an early postcard, came into American gardens by the mid-eighteenth century. The large showy scented flowers became very popular with the Dutch. *The Cottage Collection*

Bleeding Heart (*Dicentra spectabilis*) was a very popular Dutch country perennial. It had been introduced into America in the mid-nineteenth century. It was ubiquitous by 1875. *The Cottage Collection*

Nasturtium (*Tropaeolum majus*) was and remains a popular regional flower. The Dutch often pickled the flower buds, which were a caper substitute, and used the leaves as a salad green. These were photographed in a bed near the Stone Tavern. *Landis Valley Museum, Craig Benner, photographer*

Lilac (*Syringa vulgaris*) has been grown in Dutch country since the earliest days of settlement and had been grown in Europe for centuries before. It was greatly admired for its richly scented flowers. *The Cottage Collection*

The strawberry (*Fragaria vesca*) was a delicacy to the Dutch. They were not regularly cultivated until late in the nineteenth century. Henry H. and Emma usually bought theirs from neighbors. *The Cottage Collection*

The four-square garden at the Log Farm of the Landis Valley Museum is built next to the pigpen, *der Seischtall*. Any non-toxic vegetation, unfit for human consumption, was fed to the pigs. An Adam and Eve plant (*Yucca filamentosa*) grows in the center. Boards are laid in the near bed to help in planting. *Landis Valley Museum*

A blooming Adam and Eve plant (*Yucca filamentosa*) is at the center of a traditional garden. Spectacular in flower, the plant was also known as "The Lord's Candle." *Landis Valley Museum*

Watering a raised-bed garden was very arduous during a dry spell. It was, as suggested in this photograph, a chore often assigned to children. While some gardens were watered directly from buckets, watering cans were used from an early date. *Landis Valley Museum*

Broad-leaved Prickly Spinach (*Spinacia oleracea*), an old variety characterized by large, thick, deep green leaves and prickly seed, grows next to Black Seeded Simpson lettuce (*Lactuca sativa*), an old variety still widely grown (right). To the left is seen a board laid across the bed to ease cultivation. *Landis Valley Museum, Craig Benner, photographer*

Horses at the Log Farm. Horse manure was highly prized for growing flowers – especially roses. *Landis Valley Museum*

Onions grow in the foreground of the Log Farm garden. The tile roofed stone building to the left is the springhouse. It was important to have a water source close at hand to a garden. The building to the right is the pigpen, whose inhabitants converted garden waste into the manure favored for vegetable gardens. *Landis Valley Museum*

Looking as if they had stepped out of a Jacob Maentel (1778-?) painting, two young interpreters pose with a rooster in front of the garden at the Log Farm. *Landis Valley Museum*

Climbing beans grow on bean teepees typical of old German gardens. The Stone Tavern at Landis Valley Museum is in the background. *Landis Valley Museum, Craig Benner, photographer*

Nettie May Landis, as a young girl, stands outside the garden gate (*s. daerahe* or *s daerli*) typical of a German garden. The board-covered path was not. Lush tall dark colored dahlias (*Dahlia merkii*) bloom to the right *Henry K. Landis, photographer, Landis Valley Museum Archives*

Nettie May Landis and one of the Landis family dogs pose inside a somewhat weedy Lancaster County garden. Notice the Lombardy poplars (*Populus nigra cv. Italica*) in the distance. *Henry K. Landis, photographer, Landis Valley Museum Archives*

Nettie May Landis admires a snowball bush (*Viburnum opulus*) on the lawn in front of her home. The plant was known in America since the eighteenth century. A Lombardy poplar (*Populus nigra cv. Italica*) in full foliage is at the left. *Henry K. Landis, photographer, Landis Valley Museum Archives*

A "hired girl, probably a farm wife earning extra money," pushes a wheelbarrow near the kitchen garden at Landis Valley in the early spring, around the turn of the twentieth century. Rural women usually shaded themselves when they went outside. Suntans were not considered to be attractive. *Henry K. Landis, photographer, Landis Valley Museum Archives*

The Harmony Society, a communitarian religious group with roots in Württenberg in Germany, developed elaborate gardens, now being restored, at Economy, their prosperous community in present day Ambridge, in western Pennsylvania. The garden pavilion contained a wooden statue of a female called "Harmonie." Plantings are visible in front of the home of the society's founder, George Rapp (1757-1847). The community's church tower dominates the area. Old Economy is now a museum administered by the Pennsylvania Historical and Museum Commission. *Author's photograph*

Henry H. and George D. grew sunflowers (*Helianthus lactiflorus*) along the fence that stood between the wagon sheds built by the Oberlander Trust, c.1940. *Henry K. Landis, photographer, Landis Valley Museum Archives*

Henry at a fence behind the Yellow Barn holds a kitten. Morning glories, probably "Heavenly Blues" (*Ipomoea tricolor*), grow vigorously. The other side of the fence also had a vigorous growth of Hollyhocks (*Alcea rosea*). *Landis Valley Museum Archives*

George D. and Henry K. Landis' favorite garden was a colonial revival style herb garden behind the Stone Tavern. They often tended it themselves, on occasion with the help of the Lancaster Garden Club. George poses with a hoe and Henry with a watering can. A pot of petunias stands on the low stone wall surrounding the garden. An arbor, as yet unplanted, connects the Tavern to the Gun Shop. Henry sits contemplating his accomplishments. *Landis Valley Museum Archives*

The Landis brothers liked dahlias, which they scattered over their museum grounds. A stone horse-watering trough is in the middle of this view. *Henry K. Landis, photographer, Landis Valley Museum Archives*

Henry K. Landis and a guest play with kittens in the herb garden behind the Stone Tavern c.1945. *Landis Valley Museum Archives*

The Pennsylvania German love of color, combined with Milton S. Hershey's chocolate fortune, created the extraordinarily colorful Hershey Rose Garden built on a level area below his unique Hotel Hershey, designed by Hershey's favorite architect-designer D. Paul Witmer (1894-1981). The Gardens overlook the Hershey factory, the fortune's source. *The Cottage Collection*

The future generation: Landis Valley Farm Museum Director, Steve Miller, restrains his young grandson, Landis Fluck, at the small stream that runs through the farmette's former well-watered meadow area. The daylily (*Hemerocallis fulva*), an escaped old world cultivar, is a common wildflower in Central Pennsylvania. *Anne Miller, photographer*

German commercial horticulture at its most flamboyant is witnessed each autumn at Otts Exotic Greenhouse outside of Schwenksville. From atop a mountain of 10,000 chrysanthemum plants you look out on colorful growing fields and a showy sales area. *The Cottage Collection; author's photographs*

VI. Heirloom Seeds

Gude Suma, Gude Gaarde (Good Seeds, Good Garden) is a Pennsylvania German mantra. And the Pennsylvania Germans are justifiably respected for their frugality and their attention to tradition in having saved many varieties of garden plants from extinction in our time of professional hybridization. While we have records of seeds being for sale in Pennsylvania dating to the eighteenth century, it was not until after the Civil War and the rise of large commercial growers that most gardeners and farmers stopped saving their own seeds. The late nineteenth century also saw the beginning of devotion to the use of specialist produced hybrid seeds.

The genesis of the modern seed is succinctly explained in Sue Strickland's *Heirloom Vegetables.*

> …For millennia, farmers and gardeners were the plant breeders, improving crops by selection from generation to generation. In the 19th century, rather than waiting for changes to occur at random in nature, a few experimenters started to make deliberate crosses between plants. By choosing the plants carefully, these early hybridizers hoped that the offspring would have the desirable characteristics of both parents. …
>
> At first new varieties were developed by individuals and small family-based seedsmen, still with the involvement of farmers and gardeners, but techniques gradually became more sophisticated. There were also pressures to make plants of the same variety less variable, so those with slightly "wrong" characteristics (for commercial purposes) were weeded out. Over several generations varieties were developed that were very uniform and reproduced that uniformity; these were known as "pure lines." In the 1930s the first "F1 hybrid" corn varieties were produced, the seed coming from crossing two inbred pure lines. F1 hybrids of other crops soon followed.
>
> In the last 25 years [from 1998], plant breeding has become the specialist realm of microbiologists, geneticists and other scientists. It is a complex major industry. Dozens of potato varieties, for example, can be screened for blight resistance in the glasshouse within 3-4 weeks of germination, rather than growing them in a field for several months, and any useful material resulting can be held in test tubes using tissue culture. Genetic engineering allows specific genes in plants to be located and marked in the laboratory, and then moved between species in a way that was never possible with traditional hybridization methods.
>
> Today farmers and gardeners are seldom involved in plant development. It is geared to the perceived needs of industry, which is typically concerned with relatively short-term gain. Plant breeding has also become extremely expensive: for example, the development and marketing of one of the first genetically engineered varieties to be released, the Flavr Savr tomato …, is said to have cost in excess of 95 million US dollars. With such investment, it is no wonder that a few varieties dominate the shelves of superstores, and the pages of catalogs.

Hybridizer Luther Burbank (1849-1926) became a national celebrity whose feats in improving vegetables and flowers was widely applauded. He was proclaimed a genius by, among others, both Henry Ford and Thomas Alva Edison. While he worked cooperatively with many seed houses, he was especially active with Pennsylvania W. Atlee Burpee and Company, founded in 1876 by Burbank's near contemporary, Quaker descended, W. Atlee Burpee (1858-1915). Burpee's growing fields in Bucks County became famed for their Fordhook hybrids named for their farm, but many of them, the early ones, were developed by Burbank. Hybrids became the glamour crops of America. They reflected the American ethos: New! Bigger! Better! Hybrids often provide higher yields and more disease resistance than their open pollinated ancestors. The new varieties were often apparently sturdier and longer lasting than the older ones. As one wag observed, "The ideal American strawberry is as large as an Idaho potato, and just as tasty."

Much of the appeal of heirloom seeds has to do with a nostalgia for times past, as well as a concern for the tastes and aromas, now almost gone. Nettie May Landis posing in a thicket, in the late nineteenth century, epitomizes this quest to recapture the perceived innocence of an earlier time. *Landis Valley Museum Archives, Henry K. Landis, photographer*

By the twentieth century most Pennsylvania farmers and gardeners were buying their seeds and proud of their use of hybrids...this even among many of the Plain people. After Amish farmer Joseph F. Beiler's death he was eulogized for his piety and his traditional ways, but as a farmer, within community restraints, he was as progressive as he could be. "He was one of the first to use a field chopper to cut hay, blow it onto a wagon, and haul the feed to the cows. [and] New seeds and county agents' recommendations all caught his attention."

As we saw in Chapter IV, Henry H. Landis grew the latest varieties of potatoes available – and he no doubt favored up-to-date seeds as well, although none are mentioned specifically.

While modern seeds are very well adapted to commercial production, many hybrids gave their best yields only under ideal conditions. Additionally, many modern hybrids were deficient in both taste and aroma or scent from traditional varieties – which were soon almost completely unavailable, unless you personally knew a seed saver. Beginning in the 1960s, orchardists and gardeners began to realize that thousands of varieties of fruits, vegetables, and flowers were disappearing. We were not only losing their special qualities, but their precious genetic material was slipping away. Monocultures were becoming manifold and genetic diversity was in danger of being lost.

Many people were concerned about this. Kent and Diane Whealy became the most successful of people to do something about the situation. In 1972 Diane Whealy's grandfather Otto entrusted to his granddaughter and her husband the seed of three plants that had been brought from Bavaria to northeastern Iowa. "These were two vegetables—a large German tomato and a prolific climbing bean—and a beautiful dark strain of the flower Morning Glory."

Soon after Grandpa Otto's death, Kent Whealy recalls:

An interest in heirloom fruit varieties predates the heirloom seed movement. The Globe Peach and the Rutter Pear are examples of fruit varieties that originated in Pennsylvania and were grown by Dutch farmers. Both are very scarce antiques today. *The Cottage Collection*

Diane and I immediately started trying to locate other families who were also keeping heirloom seeds, hoping to increase the genetic diversity available to gardeners growing healthy food for their families (43 million US families, about 40%, grow some part of their own food, and two-thirds of the world's people live on what they can grow). We soon discovered a vast, almost unknown genetic treasure quietly being maintained by elderly gardeners and farmers. Heirloom seeds are especially prevalent in isolated mountainous areas, such as the Ozarks, Smokies, and Appalachians, and also among traditional peoples such as the Mennonites, Amish and Native Americans. It became immediately apparent that these were excelled home garden varieties, often extremely flavorful, tender, and productive. Gardeners are continuously growing and comparing varieties: those that don't measure up are quickly discarded, certainly not maintained for 150 years or more. And over such long periods, many of these heirloom seeds had slowly developed resistances to local diseases and insects, and had gradually become well adapted to specific climates and soil conditions. This heritage of heirloom seeds, which vastly outnumbers the offerings of the entire garden seed industry in North America, had never been systematically collected.

Diane and I founded the Seed Savers Exchange in 1975. For more than two decades we have continued to locate gardeners keeping heirloom food crops, and have organized them into an annual seed exchange. Each January SSE publishes a yearbook that lists all of the seeds being offered by our members. In 1975 that network consisted of 29 members offering a few dozen varieties through a six-page newsletter, now there are 1,000 'listed members' offering nearly 12,000 rare varieties through the current 400-page *Seed Savers Yearbook* sent to 8,000 gardeners. Over the years SSE's members have distributed an estimated 750,000 samples of rare seeds that were unavailable commercially and were often on the verge of extinction.

Today there are organizations devoted to seed saving in various countries and the heirloom seed phenomenon developed as many forces came together: a search for heritage, the quest for healthier foods, a widespread rejection of corporate culture and the homogeneity it fosters, and the maturity of living history museums.

Black Raspberry (*Rubus occidentalis*) is an American native bramble which was widely adopted and grown by the Pennsylvania Germans. It is available in numerous modern selections and cultivars. *M. Susan Richman, photographer*

Redbud or Judas Tree (*Cercis Canadensis*) is an American native flowering tree that is very popular with the Dutch and has been widely used as a small lawn tree. It also grows wild in many area woods, often in association with Dogwood (*Cornus florida*). *Author's photographs*

As discussed in Chapter V, the Landis brothers were not sensitive to the need for the preservation of gardens, let alone seeds. Keyser's article documented for many the need for historically correct garden forms and the garden at the Log Farm was the beginning of historically accurate gardening being introduced at Landis Valley. Volunteer Lee Stoltzfus, born Amish and today a rare books dealer specializing in garden books, along with others, started to voice concerns about the varieties of plants being grown in the garden. Then a photographer, Stoltzfus became interested in his ancestral garden forms and built his own garden along historic lines, and developed a sideline of designing gardens for others. Perhaps he first directly articulated the need for the seed program that would become the Heirloom Seed Project (HSP). There was a precedent. He knew that within the profession, there were already sources for historically correct apple varieties, and appropriate animal varieties.

One of those most concerned with plants grown in the gardens was Stephen S. Miller, then in his twenties, the farm manager and today the Landis Valley Museum Director. One of Steve's duties was to show the gardens to visitors who increasingly started to ask questions about historical authenticity. Was this the way gardens looked? Are these the plants that they grew? Steve found himself having to say, "No, but ...." a lot. He was easily persuaded to take the concept and develop it with the help of many volunteers, including Nancy Pippert, a registered nurse by profession, who had a talent for establishing the records and databases need for the program that developed around heirloom seeds. A dedicated gardener, she also worked in the gardens, first as a volunteer, then as a part-time employee of the Landis Valley Associates. She was, indeed, a "founding mother" of the Heirloom Seed Project.

Together Steve, Nancy, and others worked successfully to persuade the administration of the Landis Valley Museum and the Pennsylvania Historical and Museum Commission to establish the pioneering Heirloom Seed Project (HSP). Publicity helped launch and establish the program. The *Lancaster New Era* not only ran articles about it but also a major editorial in support. A few years later, a two-page spread about the new program appeared in the *Philadelphia Inquirer*, which was subsequently picked up by many of the other Knight-Ridder newspapers nationally. Hundreds, or perhaps thousands, of letters came from the many German settled areas all over America – some offering support, others wanting to order seeds. Founded in 1985, the project celebrated its twentieth anniversary in 2005. While the HSP enjoys official recognition from the Commonwealth, it receives no state funds; rather it is supported by the Landis Valley Associates and the work of about fifty volunteers who care for the gardens and prepare and market the seeds. The seeds have been offered through catalogs and now, increasingly, on the Internet and by direct sales at the Weathervane Shop, the museum store, which is also operated by the Landis Valley Associates.

Starting on a shoestring budget, the new project couldn't even afford padded mailing envelopes. After complaints from customers about crushed seeds, they ingeniously discovered that toilet paper and paper towel tubes made effective mailing protectors. Accordingly, they started a collection. Museum volunteers Ray and Evelyn Althouse, wintering in Florida, would put up a notice in their winter home community and arrive at Landis Valley each spring with a station wagon full of rolls – until the need subsided.

An important and popular fund raiser for the HSP is the annual Herb and Garden Faire held each May, for which volunteers raise thousands of seedlings for sale from the heirloom seed collection. The event has grown to become the largest of its class in the East and today features "... more than 90 vendors ... with herbs, historic seeds, and garden accoutrements." The Heirloom Seed Project also propagates historic roses, which it sells at the event. Each winter, Project Director Joe Schott offers a very popular tree-grafting course featuring scions from the Landis Valley Museum's collection of historic apple varieties. The course has also developed into a regional scion exchange where collectors from neighboring states, as well as Pennsylvania, come together to exchange starts of historic varieties as they learn and practice the physical techniques of grafting.

Several active youth based programs are also associated with the HSP. The Garden Gate Program works with school age children in the region to introduce historic garden information. The Landis Valley Project, which enrolls seventh grade children from Manheim Township, where the Landis Valley Museum is located, provides garden help, plus volunteer greenhouse workers and seed packers and sorters for the HSP. The area in which the Landis Valley Museum is located is rapidly suburbanizing. A few years ago Museum Director Steve Miller asked the forty-five new recruits in the Landis Valley Program, all with Dutch names like Burkholder and Herr, "How many live on a farm?" Just two raised their hands. In 1950 about forty of them would have.

By mission, the Landis Valley Heirloom Seed project "... preserves seed varieties having historical significance to the Pennsylvania Germans between the mid-1700s and 1940." The whole project, as suggested earlier, is made possible by determined volunteers who, with professional supervision, "... work with as much accuracy as possible to present precise information about our seeds, plant varieties and historical gardening techniques." Additional information is provided in the annual catalogs the HSP distributes. These small booklets are educational as well as being sales devices. There is an almost religious zeal about the Project's goals.

One of the most frequently asked questions is how do we come upon our historical information. Local research is the first step to determining the "age" of a variety. Local research consists of people and the information that they share. Our varieties have come to us with stories of "seeing this in my grandma's garden" or "my great-grandmother brought the seeds for this plant from Germany when she and my great-grandfather immigrated." We listen for physical descriptions of the seeds, plants and fruits so that when we "trial" these varieties in our experimental gardens, we have characteristics to look for and are watchful for consistencies season after season.

Diaries, old seed catalogs and almanacs, newspapers and magazines are some of the other areas that we explore in the name of research. We work with other preservationists such as William Woys Weaver, author of *Heirloom Vegetable Gardening* and Seed Savers Exchange to assist in our documentation process. Without the cooperation, sharing and common goals of other preservationists, the Heirloom Seed Project and other organizations like us could not exist.

The Heirloom Seed Project is not only a link to our past, but also to our future. Without non-hybrid seeds, such as those the Heirloom Seed Project strives to preserve, commercial seed companies and researchers could not develop the hybrid varieties that "feed the world." Without the open pollinated varieties that we offer, many people would not know the heritage that is theirs to enjoy. Without organizations like the Heirloom Seed Project, many of our "antique" varieties would be lost forever to future generations.

Tulips (*Tulipa suaveolens*) growing in a Lancaster County garden are a Pennsylvania German icon. Single red and yellow Darwin type tulips were, historically, the most common grown in the region. They are the inspiration for the designs on innumerable *fraktur* and decorated pieces. Woods tulip (*Tulipa sylvestris*), a less showy but reliably perennial form, is widely naturalized and its small yellow flowers are also well known. Tulips, as well as other widely grown spring and early summer flowers, are featured in a decorative watercolor by an amateur Schwenkfelder artist. *Author's photograph; Schwenkfelder Library and Heritage Center*

In growing heirloom seeds, care must be taken to prevent inadvertent cross pollination which, of course, produces hybrids, of often unknown parentage. Heirloom gardeners who grow more than one variety of a particular vegetable and want to save their own seeds must be cognizant of the ways in which these plants pollinate. The following "Cross Pollination Chart" is adapted from the one published in the Heirloom Seed Project Catalogs:

VARIETY	WILL CROSS WITH	SEPARATION ISOLATION METHODS
Beans *Leguminosae Family*	Perfect flowers; Self-pollinating but may occasionally be crossed by honeybees and other insects; Crossing varies depending on type of flower, population of insects, etc.	Separate white seeded varieties. Use tall interplantings to separate varieties. May use cages for bush plants. May use blossom bags for climbers
Beets *Chenopodiaceae Family*	Wind pollinated; Beet pollen is very light & can travel up to five miles. **All beets and chard will cross with each other.**	For absolute purity, isolate by 2-5 miles. Bag or cage (Min. 6 beets per cage or bag)
Carrots *Umbelliferae Family*	Insect pollinated. **All carrots will cross with each other and with Queen Anne's Lace** *(wild carrot)*	Isolate by 1/2 mile. Hand-pollination and caging
Chard, Swiss *Chenopodiaceae Family*	See beets; will cross with all chard varieties and all beet varieties	See beets
Kale *Brassicaccae Family*	Most kale and all collardswill cross with all other members of *B. Oleracea* including cabbage, broccoli, brussels sprouts, cauliflower and kohlrabi.	Grow only one variety of *B. Oleracea* per season, or isolate by one mile or cage.
Lettuce *Compositae Family*	Bee and insect pollinates;will cross with other lettuces including wild; some dispute degree of crossing.	25 feet separation of varieties.

Caging ensures seed purity. Just before flowers open, wrap seed heads with spun polyester or put cage in place. Remove cage when plants stop flowering and begin to dry.

VARIETY	WILL CROSS WITH	SEPARATION ISOLATION METHODS
Melons *Cucurbitaceae Family*	All varieties of *cucumic melo* will cross but will not cross with watermelon or other cucurbits	Isolation; Caging with introduced pollinators; Hand pollination. Grow one variety per year.
Tomatoes *Solanaceae Family*	In dispute	Cage or separate by 100 feet.
Radishes *Brassicaccae Family*	Insect pollinated and will cross with all varieties of wild and domesticated radishes. They will not cross with any other members of *Brassicaceae Family*	Separate varieties by 1/2 mile or cage; protect from wild population by removing plants
Rutabagas *Brassicaceae Family*	Self-futile: Capable of self-pollination. May be cross pollinated by insect if different varieties grown within 1 mile of each other; Will also cross with some agricultural turnips, or Fodder Turnips, and with all varieties of Winter Rape.	Separate by 1 mile or cage
Spinach *Chenopodiaceae Family*	Wind pollinated; very light pollen plant large plantings	Grow one variety per year or bag. (Spinach plants male or female; closely in wide beds for good ratio.)
Orach *Mountain Spinach*	Wind pollinated; Very light pollen	Grow one variety per year or bag. (Spinach plants male or female; plant large plantings closely in wide beds for good ratio.)
Turnips *Brassicaccae Family Brassica rapa*	Will cross with each other and Chinese cabbages, Chinese mustard and broccoli raab. They will not cross with any other Brassica species	Isolate different varieties by 1 mile
Gourds *Cucurbitacae Family*	Cross with gourds and zucchini	Isolate varieties by 1/4 to 1/2 mile; hand pollinate if grown close together
Cucumbers *Cucurbitaceae Family*	All cukes will cross with each other	Isolate varieties by 1/2 mile or hand pollinate

Before seeds are ready to be identified as Heirloom Seeds for sale, the project always grows several generations of the plants, always carefully rougeing the plants…something home gardeners who are also seed savers must do.

> Rogueing is the process of going into the garden, going through the plants, and removing any that are not true to type in order to keep the variety pure. … It is important to do this from early in the growing season right through fruiting. Ideally, plants that are showing "off traits" can be removed (rogued) before they flower and fruit to prevent unwanted cross-pollination.
>
> In order to rogue out the plants you don't want, you have to know what to look for at various stages in a plant's life. You can gather this knowledge from books on heirloom plants or by experience. You can remove any variety that does not follow your expectations in plant size, blossom color, fruit set, and size, taste, color, and so on.

Occasionally, fluxes in environmental conditions such as weather and soil can affect the size or production of fruit. It will not affect blossom color, leaf shape or fruit shape. For example, if a tomato should be round and it appears oblong, or if a purple bean is green with white stripes, you probably have a problem!

Beans are among the easier plants to rogue out. We've noticed that some older bush bean varieties throw runners now and then. They don't remain compact like bush beans usually do. With some of these older varieties, that type of vining may be part of their background, especially if everything else is all right with the bean. However, if the runners are really long and you get the distinct impression that your plant is turning into a vine rather than a bush type, it is probably best to pull those plants. When you're walking through a planting of beans, look for obviously stunted plants or those that are growing overzealously, as well as pods that are the wrong shape, size, or color. These are the ones you'll want to rogue out.

Thomas crocus (*Crocus Tomasinianus*) has been grown in Europe since the sixteenth century and can be found naturalized in the grass around many old farms. One of the few crocus varieties not eaten by squirrels, its corms are often dug up and discarded by the pests – thereby helping to spread the flowers. *Author's photograph*

Techniques of harvesting seeds vary from variety to variety. Mostly this is a simple process that just requires patience, vigilance, and minimal equipment.

For beans, peas, corn, lettuce, and ornamentals, it is simply a matter of watchful patience. Let the plant do the work for you by letting the fruit remain on the plant until the pod has dried (in the case of beans and peas) or until the dried seeds are almost ready to fall off or blow away. Particularly for beans and peas, we stress that the perfect time to harvest the seeds is when you can hold your hand under the pods and let the seed fall into it. Lighter seed heads that depend on the wind to distribute them require only a little more patience. For example, in the case of lettuce or salsify, you must be sure to check the plants daily (twice a day is a good idea) and pluck the seeds along with the white fuzz before it blows away.

For melons the process is easier but a whole lot messier! The seeds extract sugar from the fruit while still imbedded inside. Your seeds will be most viable if, after harvesting very ripe fruit, you let it sit until it begins to soften on the outside. Open up the fruit and scoop out the seeds, rinse them off and spread them on a paper towel until dry.

Cucumber and tomatoes are the most complicated because they must go through a fermentation process before you can store them. Tomatoes can be harvested as they ripen, but cucumbers should be left on the vine until they are very over ripe. The number of family and friends you are saving seeds for will determine how many seeds you will be working with. Seeds from a few fruit can go very far.

Begin by scooping the seeds encased in their gel sacs (the membrane that surrounds the seeds) into a glass jar. Here at the Heirloom Seed Project, the preferred scooping instruments are our fingers! After you scooped the seeds from the tomatoes, put a little water in the jar, stir, label the jar with the name and the date you seeded the fruit and place the jar in a cool, dry place. You can cover it lightly with a paper towel and secure with a rubber band as the fermenting process does get smelly! Do not cover tightly with a lid and do not fill the jars more than 3/4 full to allow for escaping gas. After about three days of fermenting, (this varies with the temperature of the air), a sludge will form on top of the mixture. This consists of mold, useless seeds, and bits of pulp. Good (viable) seeds sink to the bottom of the jar and useless seeds float to the top. You can stir the mixture each day to help release any good seeds caught in the sludge. The purpose of fermenting the seeds is to destroy mold and fungi spores that could harm the seed and decrease germination rates. Allowing the seeds to ferment too long encourages the protective gel sac around each seed to break down, causing the seeds to sprout.

> Once the sludge layer has formed, it is usually safe to wash the seeds. Scoop off the worst of the sludge and start draining off the rest of the liquid. As the liquid is pouring off, you're getting rid of the useless seeds that have floated to the top during fermentation. When you see that you're reaching the seeds at the bottom of the jar, stop pouring and fill the jar with clean, cold water. Allow the seeds to settle again, and pour off the liquid. Repeat until the seeds are washed and clean and the water is clean.

The next step in the saving of seed is crucial:

> Once your seeds have been harvested and processed (fermented or shelled), they should be allowed to open air dry. Place the seeds on a paper towel on a flat dish or pan (we use plant flats) and place them out of direct light in a climate-controlled place where they will have little chance of being disturbed by curious children, cats or (worst yet) mice. Be sure to label the paper towels. Gently stir them around on the paper towel each day to prevent them from sticking together.
>
> Within a few days the seeds should be dry enough to store for the winter. It is best to store seeds in an airtight container and in a cool, dry, and light-free space, such as your refrigerator. Word of caution…if you store your seeds in the refrigerator or freezer, let them warm to room temperature before you open the jar. It is best not to mix seeds from different years. Sometimes genetic crossings don't show up right away, so you want to have a way of tracing and confining a problem should one occur.

At any given time the Heirloom Seed Project is "… researching, trailing, or preserving … about 200 different varieties of vegetables, herbs, and ornamentals, and, not surprisingly, our most popular seeds are tomatoes and beans." The Heirloom Seed Project's attention to tomatoes long precedes the contemporary craze, which is an early twenty-first century phenomenon. *New York Magazine* (September 6, 2004) notes that the "Heirloom mania has so gripped the collective culinary consciousness [that] 'it would be almost daring now to serve just a beefsteak' – even the good ones that come along in August." Rutgers University's Agricultural Department is even experimenting to find the best heirlooms to grow in New Jersey. Most interesting of all, the Procacci Brothers of Philadelphia, the largest supplier of table tomatoes in America, is now promoting the "UglyRipe," a deeply fissured tomato that looks like an heirloom and has a distinctive taste – and ships well.

> The Heirloom Seed Project collects, grows, and researches heirloom tomatoes, especially potato-leaf strains, often referred to as "German type" tomatoes. These varieties are unique because their leaves are shaped like potato leaves. The fruits of these tomatoes tend to be lobed, indicating older strains. [They] are indeterminate types, which continue to set blossoms and fruit until frost. They produce long vines. Determinate varieties are compact in growth, and their fruits ripen over a short period of time.

The 2005 catalog lists eleven pink/red tomatoes and five orange/yellow ones. The reds are Amish Paste, Belgian Beauty, Black Brandywine, German Strawberry, Howard German, Oxheart, Pepper, Pink Brandywine, Red Brandywine, Reigart, and Riesentraube ("giant grape"). The orange/yellows are Golden Queen, Hartman Yellow Gooseberry, Mammoth German Gold, Pink Grapefruit (skin yellow, flesh pink), and the Yellow Brandywine. All predate 1900. Interestingly, tomatoes were one of the last of the usual vegetables to be regularly grown by the Dutch. The tomato was not commonly cultivated until the late nineteenth century.

The most popular of the tomato varieties the Heirloom Seed Project grows is the Red Brandywine, which to aficionados is the quintessential heirloom variety. Originating in Chester County, its "… flesh is juicy with terrific tomato taste." The author's personal favorites are two lesser-known varieties:

> **German Strawberry Tomato** (pre-1900) Touted as the "Ultimate Sandwich Tomato," this fruit looks like a gigantic strawberry. These uniform fruits average 10 inches in circumference and contain little juice and firm meat. Excellent flavor.
>
> – and –
>
> **Hartman Yellow Gooseberry Tomato** (pre-1900) This cherry tomato variety is a vigorous producer of 1" size fruit. It definitely beats out its modern day hybrid relatives in taste with its mildly acidic (or tart) flavor. Imagine what an interesting tomato salad these would make combined with our Riesentraube tomatoes! Best if staked.

Food writers all point to the bean as the next heirloom star. The HSP offers beans divided into four categories. Bush Beans, Lima Beans, Pole Beans, and Wax Beans. Many have colorful names. Bush beans are compact and do not need staking. The seven are Aunt Ina's Glass, Fisher, Hutterite, Jacobs Cattle, Kerley, Lizzie Baer, Mull Kidney, and Sacrament. Two that have a long local history are the especially interesting:

> **Grandma Stober's Chow Chow Bean** (pre-1850) From Lancaster County, Pennsylvania, this is the bean used in the popular Pennsylvania Dutch chow chow relish dishes. The bean is white with a

red eye and holds its shape when cooked. It is extremely prolific after maturing in ninety days. A close relative of this bean is the Soldier Bean from New England.

– and –

Mrs. Neidigh's Six Week Bean (pre-1850) This snap bush bean has been grown by a Lancaster County, Pennsylvania, family for many years. It has an excellent flavor, but does require stringing before cooking. This heavy producer has a white seed. As with most beans, it can be dried for soups.

There are four lima beans offered (including Alma's, Pennsylvania German Red, and Verna Shirk), with the star being:

Dr. Martin's Pole Limas (pre-1900) Dr. Martin's can be quite vigorous in their growth habit with vines up to twelve feet in length, but they seem to prefer climbing a trellis to the more traditional teepee poles. The pods are usually five inches in length and contain three to four large beans. There are a number of lima bean devotees at the Heirloom Seed Project who would proclaim these limas to have the best flavor in the whole world. Best planted around the beginning of June. Matures in 90-100 days.

There are four pole beans and each is a star, including the Scarlet Runner bean which is among the oldest seeds in the catalog.

Amish Gnuttle Bean (pre-1850) Dating to the 1840s, this pole bean is similar in appearance to the Gnuttle Cutshort Bean. A wonderful description of the Amish Gnuttle Bean follows: a square-ended, short, roundish, maroon-flecked, tannish-grey seed. It truly holds up to its description and adds interest, as well as taste, to the soup pot. When cooked alone, the texture is soft and creamy. An added bonus is its propensity toward high yields after maturing in ninety days.

Pretzel Bean – Vigna unguiculata; Our Pretzel Bean is not new to the catalog. However, it was brought to our attention that it is in fact a Cowpea! Plant in blocks of six to nine plants, stake with 6' poles and you will get a beautiful tall plant with lavender flowers and curly pods at the top – much like a pretzel! Can be cooked as you would string beans when harvested young.

Scarlet Runner Bean (pre-1750) This heirloom bean has a dual purpose to serve in its lifecycle. Many people grow this bean solely for the pretty red blossoms that grow on the long vines. Others take advantage of the culinary uses of the dried bean. In the dried stage, the large seed is a strikingly beautiful black bean with purple markings. Can be grown anywhere it has support and can climb, even up corn stalks! Children love to grow this one. Matures in 65-90 days.

Wren's Egg Pole Bean (pre-1825) This bean is also known as the Speckled Cranberry Bean. It is an old variety that has been grown in kitchen gardens for many years. It should be noted that pole beans are prolific and everbearing up to the first frost. This bean is no exception. Easy to shell like peas, the beans are cooked fresh or can be frozen for later use. The pods are wide, thick, and about 5" long. The bean is tan with maroon speckles and streaks when dried and it really does resemble a wren's egg.

Only one wax bean is offered, Beste Von Allen, which "...make a nice contrast when mixed and cooked with green beans."

As for other common vegetables, there are cabbage (four, with Red Drumhead dating before 1800), carrots (one, Early Scarlet Horn, pre-1710), corn (three, including Pennsylvania Dutch Butter Flavored Popcorn, pre 1885), cucumbers (two, with Gherkin, pre-1793), kale (one), leeks (one), and lettuce (three, including Deertongue, c. 1740), peas (four, including Prussian Blue, circa 1825), spinach (one) and turnip (one). Two unusual vegetables that are seldom seen are Huberschmidts ground cherry, *Physalis Pubescens,* (pre-1800), and black salsify, *Scorzonera hispanica*, "Schwarzweisel" (also pre-1800). Also offered are four melons, the oldest being citron melon *citrullus melo*, whose flesh should be candied to enhance its taste.

Gourds are in a class of their own. The ones HSP preserves are the functional varieties – "dipper" and "large bottle," both of which can be processed into household utensils or birdhouses.

The herbs grown are agrimony, American pennyroyal, basil, caraway, chives, coriander, dill, elecampane, feverfew, garlic chives, German chamomile, horehound, lovage, mugwort, parsley, summer savory, sweet cicely, and sweet marjoram. They all have histories as flavorings, medicinals, or both. Specialty plants here are red broom corn (*sorghum)* grown for its stiff tassels, Job's Tears (*coix lachrymal-Jobii*) grown for its hard shelled seeds used in making trinkets, and mole plant (*Euphoreis lathyrus*), "*Maulwarfgraut*" in Dutch.

Grown around the perimeter of the garden to deter moles, this attractive green and white striped leaf plant with heart-shaped seed pods grows up to 4' tall. It likes full sun to partial shade and will reseed itself. Not usually seen in most modern-day gardens, you will intrigue your neighbors with this unusual looking plant. GROW WITH CAUTION AS ALL PARTS OF THIS PLANT ARE TOXIC AND SHOULD NEVER BE EATEN.

The heirloom flowers are few: blackberry lily, butterfly weed, calendula, celosia (cockscomb), columbine, hollyhock, larkspur, and safflower (false saffron). Absent from the heirloom garden offerings are bulb, corn, and tuber producing plants: onions, potatoes, gladiolus, tulips, *etc*. These are eliminated because of the special storage and handling that they require.

The last category of seed the HSP sells is field crops. The familiar ones are buckwheat, Lancaster surecrop corn, flax, and rye. Manglewurtzels, a beetlike crop that produces huge roots which is used for animal feed, and spelt (*Triticum spelta*) – long used for bread and gingerbread making among the Pennsylvania Dutch are also offered.

The Heirloom Seed Project is very proud of its *Gude Sume*, all of which the determined and dedicated gardener can save for himself. But any heirloom gardener who doesn't save seed for him or herself, please remember that the Dutch have also been buying seeds for a long time. A Reading, Pennsylvania, merchant, for example, advertised in *The Chronicle of the Times* (1831):

> Fresh Garden Seeds, just received, an extensive variety of fresh and choice Garden Seeds, warranted genuine from Philadelphia, Massachusetts, and Connecticut. The space which it would be necessary to occupy in the columns of a newspaper, to insert an entire catalogue of the subscriber's excellent assortment of Seeds, is by far too large, and must be satisfied by enumerating the different sorts of Cabbage Seeds, viz:-Early York, Early Sugar loaf, Yellow Sayvoy, Drum Head, Large Drum Head, Low Dutch, (very fine) Turnip above ground, Turnip below ground, Red Dutch, Cauliflower, Brocoli, Berecol or Kale. Catalogues will very shortly be printed and circulated.

There is yet one other topic we haven't mentioned – and that is exotic plants. You see Elephant Ears (*Colocasia Araceae*) everywhere. Even on Amish and Mennonite farms. Photographs show that the Landis family grew Amaryllis. And going through Dutch Country today, you see a wide assortment of unusual plants. I asked my gardening father-in-law about this phenomenon and he remembered the plants from his pre-Depression era childhood. He told me that "…the Dutch always liked odd plants" and he recalled that they got them from "somewhere in the Midwest". And, indeed, there are many seedhouses and mail-order nurseries such as Shumway's in Randolph, Wisconsin, in business since 1860, which offers a number of botanical oddities along with its plain everyday beans, peas, and tomatoes. Today's Shumway catalog still heavily marketed in Dutch Country features an old fashioned look emblazoned on the cover with a vignette of a bearded R. H. Shumway described as "The Pioneer American Seedsman." Hyperbole? Perhaps!

The Pennsylvania Germans in their agriculture and horticulture have always represented a mixture of the old and the new, with the balances shifting with the times. On a summer day in 1914, Henry Harrison Landis noted that he counted over seventy automobiles passing the farm in a single day. At about the same time he started to lock the chicken house door at night. In the twenty-first century the culture is being diluted as more and more Dutch and their descendents leave the farmland and enter the mainstream. But many of them still garden. Attempting to balance these factors are increasing numbers of preservation, conservation, and cultural organizations and museums. May the best of the many worlds, old and new, win.

The original catalog cover for the Heirloom Seed Project Catalog was adapted from an antique pen and ink drawing in the collections of the Landis Valley Museum and highlights the seed growing heritage at Landis Valley. *Landis Valley Museum; Landis Valley Associates*

The Pennsylvania Germans prized saffron, a spice laboriously collected from an autumn crocus, *Crocus sativus*. Once harvested, the dried spice had to age for six months to a year for the full flavor to emerge. Most Dutch families had a wooden saffron cup in which to store safely and age the spice. Some of the most wonderful of these are ones decorated by Joseph Long Lehn (1798-1892) of Elizabeth Township, Lancaster County *Heritage Center of Lancaster County*

The 20th Anniversary Heirloom Seed Project catalog is as much an educational tool as a sales device. It is filled with cultural information as well as directions for saving your own seed. *Landis Valley Associates*

An array of beans is displayed for visitors to the Seed House, an adapted two-front-door Pennsylvania German home, which is headquarters of the Heirloom Seed Project. The open jar contains Jacob's Cattle Bean, a pre-1850 bush bean grown, primarily, to be dried. A preserved flower and herb bouquet provides a decorative detail. *Landis Valley Museum*

Volunteers with the Heirloom Seed Project prepare flats and pots in which they grow seedlings of heirloom varieties for sale during Spring Garden Faire, held each May at the Landis Valley Museum. They work in one of the project's hoop houses. *Lancaster Sunday News, Blaine T. Shahan, photographer*

Young mole plants (*Euphorbia lathyrus*) thrive in a growing house at the Landis Valley Museum. Often planted around the perimeter of a garden, the herb allegedly deters mole damage. Their Heirloom Seed Project's seedline can be traced to the eighteenth century. *Lancaster Sunday News, Blaine T. Shahan, photographer*

On the most deluxe level, seed chests were used to store garden seeds that were saved from year to year. Seed chests are also found in Shaker culture … minus the decoration. Joseph Long Lehn (1798-1892) of Lancaster County decorated this rare survival. *Heritage Center of Lancaster County, acquired through the generosity of the James Hale Steinman Foundation*

Not all seed savers preserved only historic treasures entrusted to them. Perhaps more started with store-bought seeds. This opened pack of poppy seeds, c.1930, was found in the estate of a dedicated seed saver from York County – along with a small tin of poppy seeds that she had saved herself – presumably as late as the 1980s. *Landis Valley Museum, Craig Benner photographer*

What gardener could resist seeds from "Good Luck Gardens in Paradise"? Not our York County seed saver. Sensitive Plant (*Mimosa pudica*) is also commonly called touch-me-not, shame plant, or humble plant. A tropical shrub often grown as a garden curiosity, its leaves close when touched. *Landis Valley Museum, Craig Benner photographer*

Seed savers used until the late twentieth century by a frugal York County gardener. Her containers included handsewn bags so old that she might have inherited them from her grandmother. Her own containers included paper envelops, a pond-fish food box, a mayonnaise jar, and a coffee can. A rye straw basket provides a backdrop. *Landis Valley Museum, Craig Benner photographer*

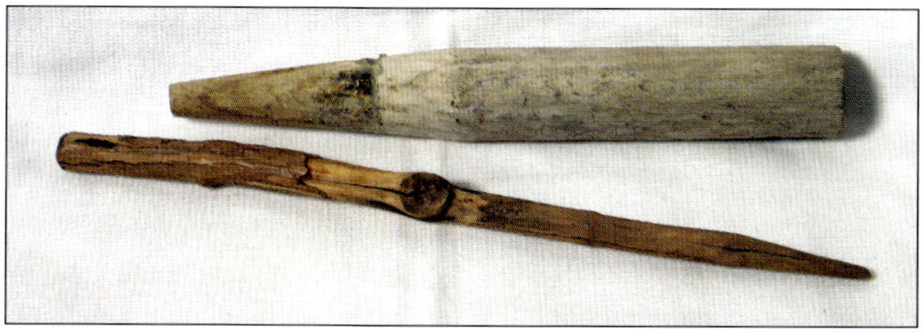

Dibble sticks were used to make indentations in which to plant seeds. The upper one is machine-made; the lower is hand-whittled. These were owned and used by a York County gardener. Many modern gardeners simply use a sharpened pencil in place of this specialized tool. *Landis Valley Museum, Craig Benner photographer*

Mexican Sunflower (*Tithonia rotundifolia*) has been a popular carefree plant for generations in Dutch country. Its large seeds are abundant and easy to collect. Because the seeds are not winter hardy, the plants never become invasive in this area. *Landis Valley Museum, Craig Benner photographer*

Acorn squash (*Cucubita pepo*) can be traced back to the Native Americans, who cultivated many forms, as did (and do) the Dutch. *Landis Valley Museum, Craig Benner photographer*

Among the many seed houses selling in the Dutch country during the nineteenth century was the German-owned Maule Seed Company of Philadelphia. Their motto was "Once Grown, Always Grown." *The Cottage Collection*

Sunflowers (*Helianthus atrorubens*) grow in a garden near the Stone Tavern. This old fashioned variety was favored for its long-blooming season and cheerful color and hues. *Landis Valley Museum, Craig Benner photographer*

The "Heavenly Blue" Morning Glory (*Ipomeo tricolor*) was introduced into the garden world in 1896 by an Ohio seed house. It subsequently became extremely popular in Dutch Country (see Chapter V). Common today, you also see it on many Amish and Mennonite farms. This specimen was raised in Cumberland County from seeds acquired during the Spring Garden Faire at the Landis Valley Museum. *William Ames LaFond, photographer*

Heirloom Tomatoes (*Lycopersicon esculentum*) are the most popular of the Heirloom Seed Project's offerings. Fruits of the Pepper Tomato, Reigart, and Red and Yellow Brandywine ripen in the growing garden. *Landis Valley Museum, Craig Benner photographer*

As a part of the process for saving tomato seeds, the seeds must go through a fermentation process before they are ready for drying and storage. Ordinary jars serve very well. *Landis Valley Museum, Craig Benner photographer*

Moon and Stars Watermelon (*Citrullus lanatas*) is an Amish favorite, which dates to 1910. Its dark skin is marked with yellow "moons" and "stars" in a completely random pattern. The pink red flesh is "wonderful good." The melon is also very popular in the American south. *Landis Valley Museum, Craig Benner photographer*

Neck Pumpkin (*Cucurbita moschata*), a vigorous grower, is favored for making pumpkin pie. Florence Steigerwalt (see Chapter I) makes an especially fine pie. *Landis Valley Museum, Craig Benner photographer*

Beans on teepees mature in one of the Heirloom Seed Project growing areas. *Landis Valley Museum, Craig Benner photographer*

Bottle gourds (*Lagenaria siceraria*) cure in one of the Heirloom Seed Project's hoop houses. The lineage of their seeds can be traced to before 1850. *Landis Valley Museum*

Lima beans (*Phaseolus lunatus*) ripen in the Heirloom Seed Project's growing fields. The perennial favorite is Dr. Martin's Pole Lima, a pre-1900 plant that is magnificently productive. *Landis Valley Museum, Craig Benner photographer*

An Heirloom Seed Project show-garden features a wide range of period flowers and herbs to engage the visitor's senses. *Landis Valley Museum, Craig Benner photographer*

A regional love of color and of exotics is seen in this contemporary suburban Lancaster County front yard, which combines Scarlet Sage (*Salvia splendens*) and banana plants (*Musa banjao*). *Author's photograph*

The recreated bank barn on the Jacob Landis farmstead is a distinctive regional backdrop for one of the Heirloom Seed Project's show gardens. *Landis Valley Museum, Craig Benner photographer*

An important fundraiser that helps support Landis Valley Museum programs is Harvest Days, held each fall. The "Pumpkin Patch" is a favorite. Although these are grown locally, alas, these are not usually grown on-site. They represent a fine crop of *Cucurbita maxima*. Most will end up as ornaments or jack-o-lanterns. *Landis Valley Museum*

Great-grandmother and great-grandson: The Landis family and Kinzers (Emma's mother's family), like many of the Dutch, kept tropical houseplants. They show up in many photographs, summering on porches. Dashing young Henry K. Landis, visiting from New York City, sits on the side porch of the family home with plants in the background (see Chapter III). Great-grandmother Anna Margaret Kinzer liked Angel Wing Begonia (*Begonia xargenteo-guttato*). Hers summers next to a vine-covered trellis. *Landis Valley Museum Archives, George Diller and Henry Kinzer Landis, photographers*

The Heirloom Seed Project and the many other programs of the Landis Valley Associates rely on volunteers who donate their time in many ways. Young Andrew Wagner and Michael Zimmerman, in pumpkin suits, take a break from their Landis Valley Float at the Denver, Lancaster County, Fair. They hope to interest ever more people in the Landis Valley tradition. *Mike Wagner, photographer*

Lancaster Surecrop Corn is a very tall growing (usually about eight feet) dent field crop also farmed for its large ears. It has an important place in agricultural history as a parent plant of many modern day corn hybrids and is a star of the Landis Valley preservation project. Leonard Schott of Cornwall Township, Lebanon County, shows off some amazing ears he grew in the 1920s. *Joe Schott Collection*

Bibliography

Adams, Denise Wiles. *Restoring American Gardens: An Encyclopedia of Heirloom Ornamental Plants, 1640-1940.* Portland, Oregon and Cambridge, U.K.: Timber Press, 2004.

Ashworth, Suzanne. *Seed to Seed: Seed Saving and Growing Techniques for Vegetable Gardeners.* 2nd ed. Decorah, Iowa: Seed Savers Exchange, 2002.

"At the Greenmarket: You Say Tomato," *New York Magazine.* September 6, 2004, 127.

Bartram, William. *Travels and Other Writings.* New York: Library of American Life, 1996.

Beam, C. Richard and Joshua R. Brown, Eds. *The Comprehensive Pennsylvania German Dictionary Vols I and II.* Millersville, Pennsylvania: Center for Pennsylvania German Studies, 2004, 2005.

Beck, Berton E. "Grain Harvesting in the Nineteenth Century." *Pennsylvania Folklife* XXIII (Summer 1974), 43-46.

Billard, Jules B. "The Revolution in American Agriculture." *National Geographic Magazine.* CXXVII (February 1970), 147-185.

Boyer, Dennis. *Once Upon a Hex: A Spiritual Ecology of the Pennsylvania Germans.* Oregon, Wisconsin: Badger Books, Inc., 2004.

Brendle, Thomas and Claude Unger. *Folk Medicine of the Pennsylvania Germans.* Breinigsville, Pennsylvania: Pennsylvania German Society, 1935.

Bucher, Robert C. "Grain in the Attic." *Pennsylvania Folklife* 13, no. 2 (Winter 1962-1963), 7-15.

Bucher, Robert C. and Alan G. Keyser. "Thatching in Pennsylvania." *Der Reggeboge* 16, no.1 (1982), 1-23.

Buffington, Albert F. et al. *Ebes Fer Alle-Ebber Ebbes Fer Dich.* Breinigsville, Pennsylvania: Pennsylvania German Society, 1980.

Cowell, F. R. *The Garden as a Fine Art: From Antiquity to Modern Times.* Boston: Houghton Mifflin Company, 1978.

Cowie, Denise. "Period Plantings." *Philadelphia Inquirer,* August 5, 2005, E1-4.

Decoteau, Randall. "Historic Harvest: Heritage Fruits and Vegetables." *New England Antiques Journal,* September 2004, 51-54.

Dillon, Clarissa. "I Like Pigs." *ALHFAM Bulletin.* XXXIV (Winter 2005), 11-12.

Earnest, Russell and Corinne. *Flying Leaves and One-Sheets.* New Castle, Delaware: Oak Knoll Books, 2005.

Emsminger, Robert F. *The Pennsylvania Barn: Its Origin, Evolution and Destruction in North America.* Baltimore: The Johns Hopkins University Press, 1992.

Fegley, H. Winslow. *Farming, Always Farming: A Photographic Essay of Rural Pennsylvania German Land and Life....* Birdsboro, Pennsylvania: The Pennsylvania German Society, 1987.

Fiorillo, Victor. "The Last Saffron Farmer." *Philadelphia Magazine* (December 2005), 213-214.

Fletcher, Stevenson Whitcomb. *Pennsylvania Agriculture and Country Life, 1640-1940,* 2 vols. Harrisburg, Pennsylvania: Pennsylvania Historical and Museum Commission, 1950.

Frederick, J. George. *Pennsylvania Dutch Cook Book.* New York: Dover Publications, Inc., 1971. Reprint of 1935 edition.

Gardner, Frank D. *Traditional American Farming Techniques.* Guilford, Connecticut: The Lyons Press, 2001.

Garvin, Beatrice B. and Charles F. Hummel. *The Pennsylvania Germans: A Celebration of their Arts.* Philadelphia, Pennsylvania: Philadelphia Museum of Art, 1983.

Gladfelter, Charles H. *Pennsylvania Germans: A Brief Account of their Influence on Pennsylvania.* University Park, Pennsylvania: Pennsylvania Historical and Museum Commission, 1990.

Goldman, Amy. *The Compleat Squash: A Passionate Grower's Guide to Pumpkins, Squashes, and Gourds.* New York: Artisan Books, 2005.

H. G. E. "Plan of a Barn Near Brownsville, Pa". *The Cultivator* V (December 1838), 165.

Hedrick, Ulysses Prentiss. *A History of Horticulture in America to 1860.* New York: Oxford University Press, 1950.

Heirloom Seed Project. *Year 2005 Catalog.* Lancaster, Pennsylvania: Landis Valley Museum, 2005.

Hollander, Stacy C. *et al. American Radiance: The Ralph Esmerian Gift to the American Folk Art Museum.* New York: Harry N. Abrams, Inc., 2001.

Hollyday, Guy. "Ephrata Cloister Wills." *Pennsylvania Folklife* XXII (Summer 1973), 11-21.

Hyams, Eduard. *A History of Gardens and Gardening.* New York and Washington: 1971.

Jellicoe, Sir Geoffrey, *et al.*, Eds. *The Oxford Companion to Gardens.* Oxford and New York City: Oxford University Press, 1986.

Johnson, Elizabeth. *Landis Valley Museum*. Harrisburg, Pennsylvania: Stackpole Books, 2000.

Johnson, Linda A. "Seeking the Best Tomatoes." *Lancaster New Era*. August 23, 2004, Business Monday, 4.

Joyce, David, Ed. *Garden Styles: an Illustrated History of Design and Tradition*. London: Pyramid Books, 1989.

Keyser, Alan G. "Gardens and Gardening Among the Pennsylvania Germans." *Pennsylvania Folklife*. XXI (Spring 1971), 2-15.

Keyser, Alan G. and Ellen Gehret. "Memories of Everyday Life in Late 19th Century Pennsylvania: Excerpts from an Interview with Hannah Rittenhouse Clemens." *Mennonite Heritage Center Quarterly* (Fall 2005), 2-4.

Klein, Georges. *Le Musée Alsacien de Strasbourg.* Strasbourg, France: Musées de Strasbourg, 1970.

Lainhoff, Thomas A. "The Buildings of Lancaster County, 1815." Unpublished Master's Paper. Pennsylvania State University at Harrisburg, February 1981.

Landis, Henry Harrison. "Diary." Manuscript Transcribed by Dawn L. Fetter, Landis Valley Museum Archives, n.d.

Landis, Henry Kinzer. "Diary." Manuscript fragment. Landis Valley Museum Archives, n.d.

Lanier, Gabrielle M. "Landscapes" in *Proceedings of Vernacular Architecture Form, Harrisburg, Pennsylvania, 2004.* Interpretive Essays Section, 9-22.

Lasansky, Jeanette. *A Good Start: The Aussteier or Dowry*. Lewisburg, Pennsylvania: Oral Traditions Project of the Union County Historical Society, 1990.

Lasansky, Jeanette. *Willow, Oak, and Rye*. University Park, Pennsylvania: Pennsylvania State University Press, 1979.

Levesque, William R. "Ugly – But Tastes Great." *Lancaster New Era*. March 4, 2005, C, 5-6.

Lewis, Arthur H. *Hex*. New York: Pocket Book, 1969.

Lick, David E. and Thomas R. Brendle. *Plant Names and Lore Among the Pennsylvania Germans*. Breinigsville, Pennsylvania: The Pennsylvania German Society, 1923.

Long, Amos, Jr. *The Pennsylvania German Family Farm.* Breinigsville, Pennsylvania: The Pennsylvania German Society, 1972.

Male, Carolyn J. *100 Heirloom Tomatoes for the American Garden*. New York: Workman Publishing, 1999.

Marranca, Bonnie, Ed. *American Garden Writing: Gleanings from Garden Lives Then and Now*. New York: Penguin Books, 1988.

Martin, Helen R. *Tillie, A Mennonite Maid: A Story of the Pennsylvania Dutch.* Reprint. Ridgewood, New Jersey: Gregg Press, 1968.

Martin, Tovah. *Heirloom Flowers: Vintage Flowers for Modern Gardens.* New York: Fireside, 1999.

Michener, James A. *Centennial*. New York: Random House, 1974.

Miller, Stephen S. "A Finer Focus of Pleasurable Living: Changes in Manheim Township, Lancaster County, Pennsylvania Agriculture, 1880-1920." Unpublished Masters Production. The Pennsylvania State University at Harrisburg, 1994.

Moss, Roger W. *Historic Houses of Philadelphia: A Tour of the Region's Museum Homes*. Philadelphia, Pennsylvania: University of Pennsylvania Press, 1998.

Moyer, Dennis K. *Fraktur Writings and Folk Art Drawings of the Schwenkfelder Library Collection*. Kutztown, Pennsylvania: The Pennsylvania German Society, 1997.

Murtagh, William J. *Moravian Architecture and Town Planning: Bethlehem, Pennsylvania, and Other Eighteenth-Century American Settlements.* Chapel Hill, North Carolina: The University of North Carolina Press, 1967.

Nizalowski, Ed. "Hogs Feeding on Mast. An Internet Inquiry." *ALHFAM Bulletin*. XXX (Winter 2005), 8-10.

Null, Hester M., Ed. *The Landis Valley Cookbook: Pennsylvania German Foods and Traditions*. Lancaster, Pennsylvania: Landis Valley Associates, 1999.

O'Hara, John. *A Rage to Live*. New York: Random House, 1949.

Partridge, Michael. *Farm Tools Through the Ages*. Boston: New York Graphic Society, 1973.

Penn, L. S. "Pennsylvania Barn, and Out-buildings." *The Cultivator* V (April 1838), 48.

Richman, Irwin. *German Architecture in America: Folk House, Your House, Bauhaus.* Atglen, Pennsylvania: Schiffer Publishing, Ltd., 2003.

Richman, Irwin. *Pennsylvania German Arts: More than Hearts, Parrots, and Tulips.* Atglen, Pennsylvania: Schiffer Publishing, Ltd., 2001.

Richman, Irwin. "The Pennsylvania-German Four Square Garden." *The Magazine Antiques*, July 2001, 92-97.

Sauers, W. Ray. "Fruit Harvesting and Preservation in Early Pennsylvania." *Pennsylvania Folklife* XXIII (Spring 1974), 38-43.

Shoemaker, Alfred A. *Christmas in Pennsylvania: A Folk Cultural Study*. Reprinted. Harrisburg: Pennsylvania: Stackpole Books, 1999.

Shoemaker, Alfred A. *Easter Tide in Pennsylvania: A Folk Cultural Study*. Kutztown, Pennsylvania: Pennsylvania German Folklife Society, 1960.

Spencer, Darrell. *The Gardens of Salem: The Landscapes History of a Moravian Town in North Carolina*. Winston-Salem, North Carolina: Old Salem, Inc., 1997.

Staff of the Liberty Hyde Bailey Hortorium. *Hortus Third: A Concise Dictionary of Plants Cultivated in the United States and Canada*. New York and London: Macmillan Publishing Company, 1976.

Stark, Les. *Hempstone Heritage I: In Accordance with their Wills: "All the Heckled Hemp She Can Spin"*. Morgantown, Pennsylvania: Hempstone Heritage, 2005.

Steigerwalt, Pansy N., comp. *The Steigerwalt Family, 1767-1979*. 3rd ed. Hazelton, Pennsylvania: Transnational Printing and Copy Center, 1980.

Stein, Sara. *Planting Noah's Garden*. Boston and New York: Houghton Mifflin Company, 1997.

Strickland, Sue. *Heirloom Vegetables: A Home Gardener's Guide to Finding and Growing Vegetables From the Past*. New York: Fireside, 1998.

Sumner, Judith. *American Household Botany: A History of Useful Plants 1620-1900*. Portland, Oregon: Timber Press, 2004.

Thompson, Ann Newlin, *et al. Germantown Green: a Living Legacy of Gardens Orchards, and Pleasure Grounds*. Philadelphia, Pennsylvania: The Wyck Association, The Germantown Historical Society, The Maxwell Mansion, 1982.

Trout, Jennifer L. "Saffran or Saff(e)rich Among the Pennsylvania Dutch of Lancaster County, Pennsylvania." *Journal of the Center for Pennsylvania German Studies,* Winter 2006, 3-7.

Van Gorden, John H. *Country Tradition*. Binghamton, New York: Homes Guild of Wyoming Conference, United Methodist Church, 1975.

Weaver, William Woys. *Country Scrapple: An American Tradition*. Mechanicsburg, Pennsylvania: Stackpole Books, 2003.

Weaver, William Woys. *Heirloom Vegetable Gardening: A Master Gardeners Guide to Planting, Seed Saving and Cultural History*. New York: Holt, 1997.

Weaver, William Woys. *Pennsylvania Dutch Country Cooking*. New York: Abbeville Press, 1993.

Weaver, William Woys. *Sauer's Herbal Cures: America's First Book of Botanic Healing, 1762-1778*. New York and London: Routledge, 2001.

Weaver, William Woys. *Sauerkraut Yankees: Pennsylvania Dutch Food & Foodways*. Mechanicsburg, Pennsylvania: Stackpole Books, 2004.

Weigel, George. "What's Old is New Again." *People Places Plants,* Mid-Atlantic Edition, Spring 2004, 28-32.

Wenger, Samuel E., compiler. *A Combined Landis/Landes Genealogy Report of the Descendants of Hans Landis and Katharina Schinz*. Akron, Pennsylvania: Samuel E. Wenger, 2005.

Weygandt, Cornelius. *The Dutch Country Folks and Treasures in the Red Hills of Pennsylvania*. New York: D. Appleton-Century Company, 1939.

Witness. Directed by Peter Weir. Paramount Films, 1985.

Yoder, Don, Ed. and trans., *The Picture-Bible of Ludwig Denig: A Pennsylvania German Emblem Book*. New York: Hudson Press, 1990.

Yoder, Don. *Discovering American Folklife: Essays on Folk Culture and the Pennsylvania Dutch*. Mechanicsburg, Pennsylvania: Stackpole Books, 2001.

Yoder, Don. *Groundhog Day*. Harrisburg, Pennsylvania: Stackpole Books, 2001.

Yoder, Don. *The Pennsylvania German Broadside*. University Park, Pennsylvania: The Pennsylvania State University Press, 2005.

Index